One Health

This collection of reflective, critical, philosophical, and practical chapters represents the author's 60 years as a veterinarian, ethologist, and bioethicist. The rising incidence of zoonotic diseases from farmed animals and wildlife in the expanding human population and so-called reverse zoonoses where humans are infecting other species are existential concerns. These concerns are linked with anthropogenic climate change and our impact on ecosystems which threaten biodiversity and the health and future of *Homo sapiens* and many other species.

These interconnected issues are examined in this book, broadening the scope and agenda of what is currently more narrowly practiced as preventive medicine. The author calls for greater emphasis on holistic preventive health-care maintenance in response to the escalating costs of human and companion animal health problems, the welfare of factory farmed animals, and the endangered status of many wild species.

Our species now faces a complex existential crisis that must be addressed in an interdisciplinary way, because there are multiple contributing factors; factors that call for the insights of science and bioethics. Fearlessly tackling contentious issues and 'wicked problems,' Dr. Michael W. Fox offers an integrated perspective of what One Health looks like on the ground.

CRC One Health One Welfare

Learning from Disease in Pets: A 'One Health' Model for Discovery
Edited by Rebecca A. Krimins

Animals, Health and Society: Health Promotion, Harm Reduction and Equity in a One Health World
Edited by Craig Stephen

One Welfare in Practice: The Role of the Veterinarian
Edited by Tanya Stephens

Climate Change and Animal Health
Edited by Craig Stephen, Colleen Duncan

Animal Welfare in a Pandemic: What Does COVID-19 Tell Us for the Future?
By John T. Hancock, Ros C. Rouse, Tim J. Craig

Global One Health and Infectious Diseases: An Interdisciplinary Practitioner's Guide
Edited by William Sander

One Health: Veterinary, Ethical, and Environmental Perspectives
By Michael W. Fox

For more information about this series, please visit https://www.routledge.com/CRC-One-Health-One-Welfare/book-series/CRCOHOW

One Health
Veterinary, Ethical, and Environmental Perspectives

Michael W. Fox

CRC Press
Taylor & Francis Group
Boca Raton London New York

CRC Press is an imprint of the
Taylor & Francis Group, an **informa** business

First edition published 2025
by CRC Press
2385 NW Executive Center Drive, Suite 320, Boca Raton FL 33431

and by CRC Press
4 Park Square, Milton Park, Abingdon, Oxon, OX14 4RN

CRC Press is an imprint of Taylor & Francis Group, LLC

ISBN: 9781032942681 (hbk)
ISBN: 9781032942599 (pbk)
ISBN: 9781003569787 (ebk)

DOI: 10.1201/9781003569787

Typeset in Times
by KnowledgeWorks Global Ltd.

To the late Deanna L. Krantz, my wife,
co-worker, and kindred spirit.

Omnia connexa – All things are connected

Contents

Acknowledgments

I am especially appreciative to have had the opportunity to communicate with millions of people through my internationally syndicated newspaper column *Animal Doctor* over the past 50-plus years and, in the process, to learn much from my readers about their animal issues and concerns. I am also grateful for the encouragement and inspiration of many people in the various fields in which I engaged, notably Dr. John Bowlby, Tavistock Clinic, London; Prof. E.C. Amoroso, Royal Veterinary College, London; Prof. Tony Palmer, Cambridge School of Veterinary Medicine, England; Dr. John Paul Scott and Dr. John L. Fuller, Jackson Laboratory, Bar Harbor, Maine; Prof. Barry Commoner, Washington University, St. Louis Missouri; Leo Bustad, dean of the College of Veterinary Medicine, Pullman, Washington; Prof. Calvin Schwabe, UC Davis California College of Veterinary Medicine; Prof. Konrad Lorenz, Altenburg, Austria; Prof. Van Rensselaer Potter, University of Wisconsin, Madison; Fr. Thomas Berry, New York, advocate of Earth/Creation-centered ethics; and Dr. Bruce Kaplan, co-founder One Health Initiative team/website, Sarasota, Florida.

My ultimate gratitude goes to the animal companions and all creatures, wild and domesticated, whom I have known and from whom I have learned so much. Also, thanks to Janet Lenius of Minneapolis, Minnesota, for editing my writings and to Alice Oven, Senior Editor, Life Science and Veterinary Medicine, Taylor & Francis Group/CRC Press, and the wonderful publishing team that produced this book.

About the Author

Michael W. Fox, born in England and now living in Golden Valley, Minnesota, USA, is a graduate of the Royal Veterinary College, London, with a PhD degree in medicine and a Doctor of Science degree in ethology/animal behavior from the University of London, England. He is a member of the Royal College of Veterinary Surgeons, the British Veterinary Association, an Honor Roll Member of the American Veterinary Medical Association, a long-term member of the American Holistic Veterinary Medical Association and a founding member of the International Association for Applied Animal Ethology. He helped make animal behavior and animal welfare science part of the veterinary teaching curriculum around the world, and more generally awakening public sensibility through his internationally syndicated newspaper column *Animal Doctor* both in the United States and in many other countries. He has authored over 40 books, including award-winning children's fiction. His website is www.drfox-onehealth.com.

SOME OTHER BOOKS BY THE AUTHOR

For All Our Relations: Poetic Visions and Vexations of a Veterinarian. (2024). Austin Macauley Publishers, London. U.K.

Healing Animals and the Vision of One Health. (2011). CreateSpace Publ/ Amazon.com

Animals & Nature First (2011). CreateSpace Publ./Amazon.com

Not Fit for a Dog: The Truth About Manufactured Dog and Cat Foods (2008). co-authored with Drs. Elizabeth Hodgkins and Marion E. Smart. Quill Driver Books, Sanger, CA.

Dog Body, Dog Mind: Exploring Your Dog's Consciousness and Total Well-Being (2007). The Lyons Press, Guilford, CT.

SOME BOOKS EDITED

Advances in Animal Welfare Science 1984, 1984/85 and 1986/87 with L.D. Mickley. Martinus Nijhoff Boston/Dordrecht.

The Wild Canids (reprint edition 1983 of 1975 edition published by Van Nostrand Reinhold Co. NY.). Krieger Publishing Co., Melbourne, FL.

On The Fifth Day: Animal Rights and Human Ethics (1977). Acropolis Press, Washington, DC (with R.K. Morris).

SOME CHILDREN'S BOOKS

Animals Have Rights Too (1991). Crossroads/Continuum, NY.

The Way of the Dolphin (1981). Acropolis Books, Washington, DC.

The Touchlings (1981). Acropolis Books, Washington, DC.

Ramu and Chennai [*] (1975). Coward, McCann & Geoghegan, NY.

Sundance Coyote [**] (1974). Coward, McCann & Geoghegan, NY.

The Wolf [***] (1973) Coward, McCann & Geoghegan, NY.

---

[*] Best Science Book Award, National Teachers' Association
[**] Nominee for Mark Twain Awards
[***] Christopher Award for Children's Literature

Introduction

I, who was born to die shall live. That the world of animals, and the world of men may come together, I shall live.

–Inuit Legend

This collection of reflective, critical, philosophical, and practical essays presented in the following chapters represent my 60 years as a veterinarian and ethologist (animal behavioral scientist) and bioethicist, dedicated to improving our relationships with all creatures great and small, and with nature, the natural environment that sustains us all. Also, writing the weekly syndicated newspaper column, *Animal Doctor*, for over 50 years has given me a broad appreciation for readers' concerns about animals, the health, and behavioral problems of companion animals, as well as the kinds of diagnoses and treatments they receive from veterinarians. The escalating costs of human and companion animal health problems call for greater emphasis on the holistic and integrated preventive health-care maintenance that is reviewed in this book.

The rising incidence of zoonotic diseases from farmed animals and wildlife in the expanding human population and so-called reverse zoonoses, where humans are infecting other species, are existential concerns. These concerns are linked with anthropogenic climate change and our impact on ecosystems (including agricultural soils producing nutrient-deficient and chemically contaminated crops), which threaten biodiversity and the health and future of *Homo sapiens* and many other species. These interconnected issues are examined in this book, including the importance of protecting apex predators such as the wolf, broadening the scope and agenda of what is currently more narrowly perceived and practiced as preventive medicine.

The ecological and environmental values/services of natural biodiversity are the cornerstones of public health and even our existence, as revealed by the biophysical sciences. It is in our best interests to link this knowledge with how we treat others, human and non-human, and the environment and make immediate improvements under the global bioethical banner of One Health. This is preferable, as this book demonstrates, to our continuing to exploit life, to genetically manipulate and make advances in synthetic biology with artificial intelligence enhancement to profit from marketing purported panaceas to improve the human condition.

Those around the world whose livelihoods depend on animals, those who use them for various purposes, and the millions who have come to regard them as family members, I hope to inform and inspire with my chapters reviewing what is known about animal sentience. These chapters demonstrate what we can all learn from fellow creatures and how we can assess their quality of life under our dominion. Under the banner of One Health, our own well-being and caring for animals and the environment are interdependent and call for understanding and reverential respect for all life. In the process, our own security and prosperity in body, mind, and spirit will be better assured.

Our species now faces a complex existential crisis that must be addressed in an interdisciplinary way, because there are multiple contributing factors – factors that call for the wisdom of science and bioethics to inform and inspire a United Environmental Nations.

Dr. Albert Schweitzer's advocacy of reverence for life, resonating with the spirituality and ecological sensibilities of the Anishinaabe (Native American Indians), as I detail in this book, provide the bioethical basis for One Ethic as the complementary element of One Health science and praxis. This is the One Ethic of equal consideration and justice for all sentient beings. Physician Dr. Schweitzer declared that 'Ethics is nothing other than Reverence for Life. Reverence for Life affords me my fundamental principle of morality, namely, that good consists in maintaining, assisting, and enhancing life, and to destroy, to harm or to hinder life is evil.'

Reverential respect encompasses the well-being of person and planet, species, and ecosystems. I have dedicated several chapters to various animal industries and our historical and continued and, until recently, unquestioned exploitation of fellow sentient beings. Our ultimate well-being is in adopting the vision and praxis of One Health that we are all capable of realizing and actualizing as rational, responsible citizens of planet Earth.

The One Health paradigm calls for the examination of the roots of our relationships and responsibilities for animals and the environment, including the plant and microbial life forms that sustain ecosystems aquatic and terrestrial. This calls for an interdisciplinary approach (as depicted in Figure 0.1) in which

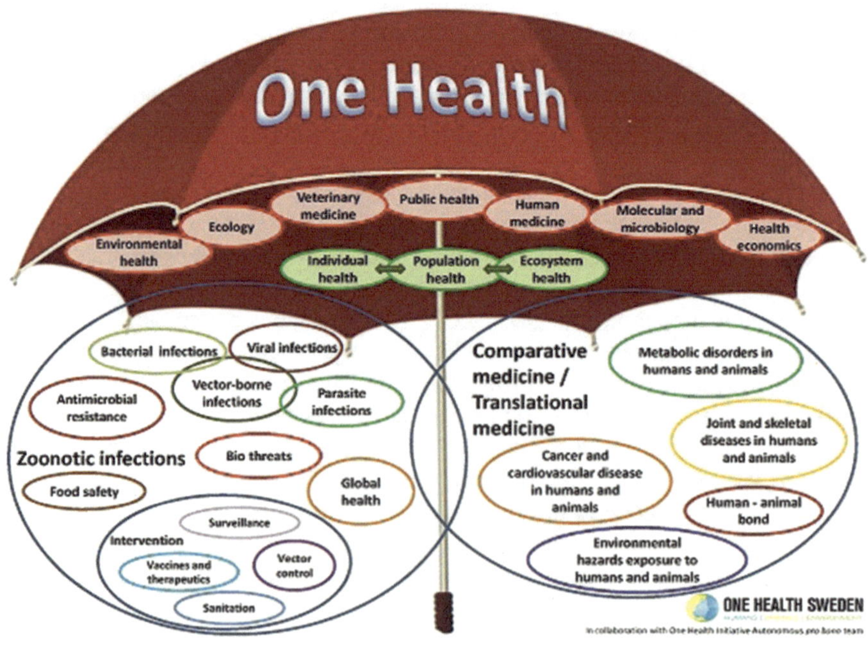

FIGURE 0.1

the veterinary profession plays a central role. Veterinarians have the most immediate, professional authority and public recognition to address those key relationships promoting health and well-being between society and animals, as well as a long association with various animal sectors and public health.

The contributions of veterinary and allied biologists, bioethicists, and human health-care professionals addressing the complexities of One Health from the holistic perspective of planetary CPR – conservation, protection, and restoration – have yet to be fully integrated and implemented. I hope this book facilitates this process. One Health for all!

Michael W. Fox, B.Vet.Med, PhD, DSc, MRCVS
Golden Valley, Minnesota

1 Recognizing Sentience to Better Protect Living Beings

Reviewing what is known about the sentience of other living beings is a prerequisite to discussing our responsibilities and relationships with them, their quality of life, and ecological services.

The scientific determination of animals' sentience – what they sense, feel, and know – is a challenge calling for an interdisciplinary application of ethology and the neurosciences. It provides an objective basis for their welfare assessment and provisions by caregivers and is the foundation of animal welfare science and associated animal protection laws.

The sentience of plants, long felt by empathic persons engaging in 'forest bathing' and the shamanic practices of indigenous peoples finding healing plants, like animal sentience, has been long denied or never even considered. This is now changing.

Indigenous peoples believed that their crops were healthier and more productive when they sang or played music for them. Nobel Prize winner and plant geneticist Barbara McClintock observed, 'For instance … if you pinch a leaf of a plant you set off [an] electrical impulse. You can't touch a plant without setting off an electrical impulse.… There is no question that plants have all kinds of sensitivities. They do a lot of responding to an environment. They can do almost anything you can think of' (1).

In the 1901 edition of *Erewhon*, Samuel Butler added a chapter on vegetable rights that reiterates Darwin's findings and parses their implications: If plants truly are conscious, intelligent beings, then how might human authors narrate their presence? And how should this new understanding of vegetal being affect humans' interactions with plants? For an interesting discussion on this subject, see the review article by philosopher Mary Bowden, 'Vegetal Being,' in Samuel Butler's *Erewhon*: *The Narrative Challenge of Nineteenth-Century Plant Science* (2).

The Secret Life of Plants by Peter Tompkins and Christopher Bird documents controversial experiments that claim to reveal unusual phenomena regarding plants, such as plant sentience, discovered through experimentation. This set the stage for more research into the sensitivities and intelligence of plants which are attuned to each other and to insects and other animals in their environments (3). Some flowers, for instance, open to make nectar more accessible in response to the sounds that bees make while hovering over them.

DOI: 10.1201/9781003569787-1

Suzanne Simard's seminal treatise *Finding the Mother Tree: Discovering the Wisdom of the Forest* (4) opens the reality of the symbiotic communication between trees and fungi and the evident sentience of interdependent plant life. Plants use chemical and electrical signals to communicate with each other. For instance, plants can communicate with insects as well, sending airborne messages that act as distress signals to predatory insects that kill herbivorous insects. Maize attacked by beet armyworms releases a cloud of volatile chemicals that attracts wasps to lay eggs in the caterpillars' bodies.

It is a matter of recorded history that the science of animal sentience, which supported the animal rights philosophy, was opposed as being unscientific and anthropomorphic. For instance, Stephen Budiansky's book, *If a Lion Could Talk: Animal Intelligence and the Evolution of Consciousness*, featured this promotional statement: 'Although Budiansky concedes that animals most likely experience emotions, he denies them consciousness, which, in his view, is inseparably linked to language, an exclusively human invention. Furthermore, Budiansky contends, animals don't really suffer, at least not the way we do, because their sensation of pain lacks a social context. Budiansky, a science writer (*The Nature of Horses*) and *U.S. News & World Report* deputy editor, uses this debatable thesis to bash the animal rights and deep-ecology movements. Whatever one thinks of the correctness of his argument, it has value as a level-headed critique of our tendency to anthropomorphize animal behavior' (5).

Budiansky was not alone in politicizing and twisting science to discredit animal welfare and rights advocacy and environmental protection. But for me, the icing on the cake was his 1996 article in *U.S. News & World Report* in which he wrote 'a surprising number of leading academic writers on animal rights and the environment share the Unabomber's paranoid hostility to science – The first is the Unabomber: [FBI's most wanted fugitive, Theodore Kaczynski] the second is Michael W. Fox, a widely published writer on animal rights and the environment' (6).

This was all part of a concerted effort to dismiss animals as sentient, conscious beings on the grounds of anthropomorphizing (having emotional states like us) to protect various animal industries and culturally accepted norms of exploitation. Instead, it was argued that animals are unfeeling automatons governed by instinctual reflexes. This 'mechanomorphizing' of other animals helped create the 'great divide,' as I call it. The exploiters of animals needed this to continue their activities in all good conscience and for protection from public accountability and censure. Distancing us from the environment and from other animals and severing empathy along with compassion was the aim to maintain the status quo of animal and environmental exploitation. Around that time, I was testifying before a U.S. Senate subcommittee to improve conditions for veal calves being raised in narrow crates in which they could not even turn around. One defendant of the industry, animal science Prof. Stanley Curtis from the University of Illinois, when asked if veal calves need to turn around replied, 'We need to do more research before we can be really sure.'

This great divide was eventually bridged by ethologists and other biological and animal welfare scientists, marked by the 2012 Cambridge Declaration on Animal Consciousness, a formal acknowledgment that many non-human animals, including mammals, birds, and cephalopods, also possess 'the neurological substrates that generate consciousness' (7).

Well before this declaration, veterinarians had been diagnosing and treating emotional conditions in animals also seen in humans, and with similar medications as for depression and separation anxiety, a field of veterinary care I helped establish in the early 1970s (8). Evidence of homologous and analogous human emotional responses in other animals, including insects, crabs, and lobsters, have been documented (9) and ethology applied to bridge our lack of empathy for invertebrate species (10).

Government animal welfare acts in the United States and other countries to establish standards that mandate humane treatment of various species, and their

FIGURE 1.1 The red X mark shows a broiler chicken fully conscious going into the scalding vat, demonstrating the industry's disregard for animal pain.

enforcement are, at best, inconsistent and need considerable revision and greater international accord. The U.S. Department of Agriculture's (USDA) Animal and Plant Health Inspection Service (APHIS) conducts annual inspections of all licensed facilities and generally oversees compliance, but many animal species are excluded from protection.

Enacted January 23, 2002, Title X, Subtitle D of the U.S. Farm Security and Rural Investment Act changed the definition of 'animal' in the Animal Welfare Act (administered by the USDA), specifically excluding birds, rats of the genus *Rattus*, and mice of the genus *Mus*, bred for use in research. Such exclusions essentially deny that these species are sentient, can experience pain and fear, and can therefore suffer. This selective speciesism is biologically absurd and scientifically and ethically untenable. If fish could scream, would that make any difference since they have no protection under such laws in most countries?

The British Veterinary Association Council has declared that crabs, lobsters, octopuses, and squid 'should be regarded in legislation as having consciousness and the capacity to experience feelings such as pleasure and pain' (11). Now this legislation has been passed into the law of the land: 'Octopuses, crabs and lobsters can experience pain or suffering, according to a review commissioned by the UK government, which has added the creatures to a list of sentient beings to be given protection under new animal welfare laws. The report by experts at the London School of Economics looked at 300 scientific studies to evaluate evidence of sentience, and they concluded that cephalopods (such as octopuses, squid, and cuttlefish) and decapods (such as crabs, lobsters, and crayfish) should be treated as sentient beings' (12).

Ethical and legal constraints on our collective negative impacts on the natural environment, biodiversity, and animal well-being are long overdue. Clearly, the concept of One Health will not advance significantly without a complementary One Ethic template: Specifically, the bioethical principle of equalitarianism (13). This principle extends the Golden Rule to embrace all sentient beings, which Albert Schweitzer advocated in his philosophy of reverence for life (14). Neuroscientist David Anderson in his book, *The Nature of the Beast: How Emotions Guide Us*, describes research from his lab that suggests the brain circuits underlying human emotions have a lot in common with circuits found in mice and even fruit flies (15).

Scientists have done many studies to determine what animals, vertebrate and invertebrate, sense and feel and think – so-called cognitive ethology. And this is as much a challenge as it is to know what another person is thinking and feeling. There are gradients of sentience, varying degrees, and kinds of sensing, feeling, intelligence, and self-awareness both within and between species, as we know from our own interactions with people and other animals in our lives. If we had no empathy, would we have any real understanding?

Scientists are elevating our understanding and appreciation for insect life, as documented in the excellent review 'Why Insects Are More Sensitive Than They Seem' by Zaria Gorvett, who writes, 'For decades, the idea that insects have feelings was considered a heretical joke – but as the evidence piles up, scientists are rapidly reconsidering' (16). Bumblebees play, according to new research led by the Queen Mary University of London published in *Animal Behaviour.* It is the first time that object play behavior has been shown in an insect, adding to mounting evidence that bees may experience positive feelings (17).

Scientists have revealed the release of endogenous opiates (enkephalins and endorphins) in many species, including earthworms (18) and fish, when subjected to traumatic injury (19). This does not mean that it is acceptable to hook fish and mutilate sentient animals because they may seem not to experience pain, but rather acknowledge that they do react to injuries and trauma as we do and should not be harmed. And the spread of sentience goes from invertebrates to mammals, from earthworms to humans.

This endogenous opiate neurohumoral system, along with oxytocin and other neurochemicals, plays a role in empathically and vicariously experiencing and responding to another's suffering (20). It is evidenced in the behavior of social animals responding to the plight of conspecifics and in their care and protection of their offspring. In their review on this subject, J. Panksepp and J.B. Panksepp state: 'Cross-species evolutionary approaches to understanding the neural circuitry of emotional 'contagion' or 'resonance' between nearby animals, together with the underlying neurochemistry, may help to clarify the origins of human empathy' (21).

Objectification and 'mechanomorphization' of other sentient beings are scientifically invalid and ethically unacceptable. One of the first contemporary scientists to confront this cultural attitude was zoology professor Donald R. Griffin in his seminal text, *The Question of Animal Awareness* (22). Scientists at the Research Institute for Farm Animal Biology in Germany are exploring the mental and emotional capacities of livestock, and they have found that pigs exhibit signs of empathy, goats have social intelligence on par with that of dogs in some cases, and cows are self-aware. These and other findings may shed light on human cognition and emotion and could change animal husbandry practices and lead to better treatment, long overdue (23).

The empathic question is not do sentient beings 'know' (and are self-conscious), but rather, can they be harmed? All sentient beings avoid being harmed, often in ways which we are only just discovering, like bacteria mutating and evolving resistance to antibiotics and trees signaling chemically to others when they are under attack by insects.

Will and instinct are interdependent, as are the senses, sensibility, and consciousness, all of which can vary as greatly within species as between species. In several realms of the senses many species are far more evolved than we, while others, as we are discovering, are emotionally more like us than they are different in their responsiveness, joys, and suffering.

ANIMAL COGNITION

Our understanding of animal cognition, of what our animal companions know and what they think, is enhanced when we can put ourselves in their place and minds. This is not a difficult process since we share similar physical body and mental brain functions, sensations, and inner experiences and desires that are expressed behaviorally in various emotional displays, vocal signals, and specific actions and reactions. We are not the only animal species with the ability to recognize another's intentions – or at least conceive of them. This is a basic component of the theory of mind, namely, the ability to attribute mental states to oneself and others, which may be facilitated by so-called 'mirror' neurons (24). Scientists S. Acharya and S. Shukla conclude that 'Essentially, mirror neurons respond to actions that we observe in others. The interesting part is that mirror neurons fire in the same way when we recreate that action ourselves. Apart from imitation, they are responsible for myriad of other sophisticated human behavior and thought processes. Defects in the mirror neuron system are being linked to disorders like autism. This review is a brief introduction to the neurons that shaped our civilization' (25).

According to one review of this subject by Jessica Marshall (26),

> Just over 20 years ago, an Italian team reported findings on macaques showing the existence of a 'surprising new class' of neurons in a particular region of the premotor cortex of the macaque brain. These neurons were active not only when the macaque performed an action – like grabbing an object – but also when the macaque watched the same action being performed by a person or another monkey. The researchers named these neurons 'mirror neurons' and proposed that they provide the basis for what became known as 'action understanding' in macaques, allowing them to interpret the intentions or goals of the person or monkey whose actions they are observing. The only other animal besides macaques and humans where mirror neurons have been found is in songbirds, says Vittorio Gallese of the University of Parma, Italy, one of the authors of the paper that first reported the neurons' existence, but he believes it is only for lack of looking.

Animal science professor Temple Grandin has proposed that animals are primarily visualizers, thinking in pictures. Grandin suggests that people such as she with autism are like animals, as they 'see, feel and think in remarkably similar ways' (27). But cognitive processes are highly complex, and reactions involve multisensory, multimodality integration (hearing, seeing, smelling, tasting, touching, etc.), which are coupled with innate and learned/conditioned emotional and neuroendocrine responses. It is too simplistic to infer, therefore, that animals see in pictures.

Adapting to domestication should not lower animals' quality of life physically or emotionally. Some animals, like humans, can suffer from attention deficit–hyperactivity disorder, a condition that has genetic, developmental, and nutritional elements in its genesis. Other animals can vary in their abilities of concentration; self-control/inhibition; learning/memory recall; and reasoning, insightfulness, and affective or emotive scope of expression, comprehension, playfulness, and empathy. Such individual variations in abilities and motivation mean there is no single way to raise and relate to other sentient beings. We must relate to each as an individual, and above all, we should not use punishment (negative reinforcement) to inhibit most instinctual reactions since they are part of every animal's feral spirit, but rather apply ways to redirect and sublimate.

Learning what animals express and how they express themselves and understand each other opens us to their realm of emotional intelligence, an invitation that is acceptable to those people who neither lack empathy and emotional intelligence nor believe that dogs, cats, and other animals are simply instinctual automations devoid of feeling and self-awareness.

SELF-AWARENESS IN ANIMALS

Some animal psychologists and others contend that the mirror test, where animals recognize that a mark has been placed on their faces, is the cardinal sign of self-awareness. Species who do react to such changes in their mirror image include humans and the great apes, dolphins, elephants, and magpies. Ethologist Jonathan Balcombe in his book *Super Fly* cites research indicating that some ant species also pass the mirror test. He asserts, 'Whatever justified angst and antipathy we may feel for certain flies, we can simultaneously cultivate respect, even reverence, for their essential place in the world' (28).

I have long contended that self-awareness in various species cannot be decided simply based on this mirror test. Many people living with cats and dogs will attest to the fact that after initially reacting to their mirror image, often as though it were another animal, their animals simply ignore their reflections. This is evidence to me that they know it is not another animal but themselves that they are seeing. Some cats will play games, stalk their image, and look behind the mirror.

ANIMAL FEELINGS AND FEELING FOR ANIMALS

In my doctoral dissertation in ethology/animal behavior, I documented emotional states in the behavior of wolves and other wild canids published in my book *The Behavior of Wolves, Dogs and Related Canids*, which we humans share (29). These include caregiving, empathy, protectiveness, affection, joyful playfulness, curiosity, fear, anxiety, and depression. So how can we continue to be indifferent to other species who share a similar emotional intelligence and degree of sentience?

This disconnect of empathy is rooted in ignorance, rationalization, denial, and beliefs so that trapping, snaring, shooting, poisoning, and wildlife killing contests (visit www.projectcoyote.org for details) continue in many countries. English philosopher Richard Ryder attributes such wanton cruelty to 'speciesism,' which shares the same root of psychopathy with racism and sexism (30). Without respect and empathy for all life, the rule of law will remain arbitrary, capricious, and self-serving. The 'great divide' separating humans from other sentient beings that were seen as living machines was promoted by French philosopher René Descartes in 1637 in his book *Discourse on the Method of Rightly Conducting One's Reason and of Seeking Truth in the Sciences* (31). This had an enduring influence on how animals were regarded and treated until attitudes began to change approximately 235 years later after English biologist/naturalist Charles Darwin published his book *The Expression of the Emotions in Man and Animals* in 1872 (32).

GRIEF AND MOURNING IN ANIMALS

Long dismissed as unscientific anthropomorphizing, the belief that many animal species exhibit signs of grief and mourn the loss of a companion has been confirmed recently by ethologists and others closely observing an animal's behavior. David Alderton's popular book *Animal Grief: How Animals Mourn* (33) provides many documented cases in several species, including Edinburgh's famed dog, Greyfriars Bobby, who spent 14 years beside his master's grave.

Asian elephants have been documented burying their dead calves with their legs poking out of the ground. The calves were 1 year old or less and were transported to premade burials of sorts – irrigation drains on tea estates in the eastern Himalayan floodplains of northern West Bengal – by herd members before being placed in holes and covered in soil (34). Researchers have previously observed African forest elephants (*Loxodonta cyclotis*) performing burial practices in which herd members cover their deceased companions and family members with branches and leaves. The new study provides the first record of Asian elephant burials and is the first known case of Asian elephants using soil in burials as well as specifically positioning the body. The report by Kaswan and Roy includes references to thanatology studies in other species such as gorillas and baboons and how they respond to the death of conspecifics. Many species clearly mourn the death of offspring and adult members of their communities.

ANIMAL SELF-HEALING AND MUTUAL AID

There is a growing body of scientific literature on the innate and acquired ability of many species to use natural ways of warding off noxious insects and of healing when they are ill, nutrient-deficient, or injured, as well as to help each other. This is yet another aspect of animal sentience that makes us appreciate their intelligence and has even helped us discover the health benefits of various botanicals. In her book *Wild Health*, author Cindy Engel provides detailed documentation of how wild animals find ways to maintain their health via these natural means (35). Anthropological studies document how humans have learned from observing animals the medical and nutritive benefits of various plants (36).

Another dimension of animal sentience is in the symbiotic relationships between species, referred to as mutualism, that help each other avoid predators, find food, and improve their health as per the little fish that clean the bodies of manatees and other large fish and insectivorous birds that remove ticks from antelopes (37). New instances are being frequently reported by field biologists that reveal the complexities of interdependence and support, as proposed by Count Peter Kropotkin in his seminal 1902 book, *Mutual Aid: A Factor in Evolution* (38).

FIGURE 1.2 Compatible cats care for each other and are generally healthier than those living alone.

ANIMAL PLAY

Play behavior seen in many species has been discussed on many levels, seeking to avoid anthropomorphism, and striving to identify purpose. In his book *Kingdom of Play: What Ball-Bouncing Octopuses, Belly-Flopping Monkeys, and Mud-Sliding Elephants Reveal about Life Itself* (39), author David Toomey explores the research into this elusive subject. At least 30 hypotheses of animal play are being investigated, though Toomey notes that 'some are little more than notions.' He finds a definition by Gordon Burghardt, an ethologist and evolutionary biologist, particularly helpful: Play is 'behavior that is nonfunctional, voluntary, characterized by repeated but varied movements, and occurring only when the animal is healthy, safe, and well-fed.' Having observed play behavior in many species, it is evident to me that the pleasure principle is evident as a self-rewarding behavior in manipulating inanimate objects; bonding behavior with playmates, probably with the release of oxytocin; learning self-control and empathy to play-fight gently; and refining motor skills for hunting, escape, self-defense, and courtship.

During play, many species show what can best be interpreted as teasing, but where there is bullying, an adult may intervene, as I have witnessed in wolves and dogs. Similar socializing and developmental and cognitive benefits of social play are evident in children whom we can 'zoomorphize' in seeing that their motivation to play is shared by many other species, eloquently detailed by Toomey. Seeing different species playing together, such as ravens and wolves, baboons and zebras, and children and their dogs, opens the doors of our perception and realization of pleasurable communion and joy in the animal kingdom (40). Deception ('lying') is also evident in some species during play and especially while hunting and when avoiding predators.

Highly empathic people involved in caregiving – doctors, spouses, parents, nurses, and veterinarians in particular – sometimes feel so deeply for their patients that they experience the other's pain, even to the point of body location. I term this 'sympathetic resonance,' which can be stressful and lead to caregiver fatigue and burnout. Similarly, highly empathic animals can become caregivers, as Alaska wolf biologist, the late Dr. Gordon Haber, detailed to me when I visited him at the foot of Mount McKinley: He saw members of a wolf pack bringing food to their pack leader/father who was injured in an elk hunt until he fully recovered.

After approximately 50 years communicating with pet owners via my weekly syndicated *Animal Doctor* newspaper column, I can attest to the empathy evident in many cats and dogs, cats lying and purring where their human companion is hurting, and dogs even scenting cancer and evidencing distress. Many dogs show their empathic understanding as guides for the blind, sentinels for diabetics and epileptics, and their presence even facilitating children learning how to read (41).

I am happy to end this chapter with the widely publicized coalition of scientists publishing a declaration that asserts: 'When there is a realistic possibility of conscious experience in an animal, it is irresponsible to ignore that possibility in decisions affecting that animal' (42). Furthermore, there is international discussion of insect sentience because billions of insects are being farmed, not just for silk but as a sustainable nutrient resource and food supplement already approved for human consumption (43,44). The discovery that ants perform surgery to remove damaged appendages of their colony members to prevent spread of infection and save lives is another revelation of sentience in insects (45). Florida carpenter ants (*Camponotus floridanus*) bite off injured nest mates' limbs to save them from deadly infections. This is the first example of animals other than humans performing such lifesaving amputations. 'The ant presents its injured leg and calmly sits there while another ant gnaws it off,' explains animal ecologist and study co-author Erik Frank. 'As soon as the leg drops off, the ant presents the newly amputated wound and the other ant finishes the job by cleaning it.'

Another indicator of sentience is in the finding that just like humans, dolphins and elephants use specific sounds to call particular individuals, such 'naming' now being reported in marmosets (46).

All creatures have intrinsic value and interests which call for objective assessment of the quality of life they have under our dominion. This is discussed in the next chapter, followed by a review of their and extrinsic ecological value, biodiversity, and their collective contribution to One Health.

ADDITIONAL NOTES

- Based on several scientific studies (Review of the Evidence of Sentience in Cephalopod Molluscs and Decapod Crustaceans, by Jonathan Birch et al, November 2021, https://bit.ly/3DUgeTe), the UK government is extending coverage under the Animal Welfare (Sentience) Bill to include cephalopods and decapods (octopuses, crabs, and lobsters).

- Ethologist Marc Bekoff (https://www.psychologytoday.com/us/blog/animal-emotions/202112/spain-joins-other-nations-in-declaring-animals-are-sentient) notes: 'As of November 2019, 32 countries have formally recognized non-human animal sentience. These are: Austria, Australia, Belgium, Bulgaria, Chile, Croatia, Cyprus, the Czech Republic, Denmark, Estonia, Finland, France, Germany, Greece, Hungary, Ireland, Italy, Latvia, Lithuania, Luxembourg, Malta, New Zealand, the Netherlands, Poland, Portugal, Romania, Spain, Slovakia, Slovenia, Sweden, Switzerland, and the United Kingdom.'

- *See also* Dana L.M. Campbell and Caroline Lee. A perspective on strategic enrichment for brain development: Is this the key to animal happiness? *Front Vet Sci*. 2021; 8: 720422. Mollie A. Bloomsmith et al. Behavioral management programs to promote laboratory animal welfare. Chapter 5 in *Management of Animal Care and Use Programs in Research, Education, and Testing*. 2nd edition. Robert H. Weichbrod et al, eds, Boca Raton (FL): CRC Press/Taylor & Francis; 2018. Helen S. Proctor, Gemma Carder, and Amelia R. Cornish. Searching for animal sentience: A systematic review of the scientific literature. *Animals (Basel)*. 2013 Sep; 3(3): 882–906.

REFERENCES

1. McClintock, B. Quoted in Ritzer, George and Smart, Barry, eds. *Handbook of Social Theory*. London, Sage Publications Ltd, 2001. p. 532.
2. Bowden, M. Vegetal Being in Samuel Butler's *Erewhon*: The Narrative Challenge of Nineteenth-Century Plant Science. *ISLE: Interdisciplinary Studies in Literature and Environment*, Autumn 2021; 28(3): 966–985. https://doi.org/10.1093/isle/isaa185
3. Tompkins, P., Bird, C. *The Secret Life of Plants*. New York, Harper & Row, 1973.
4. Simard, S. *Finding the Mother Tree: Discovering the Wisdom of the Forest*. New York, Random House, 2022.
5. Budiansky, S. *If a Lion Could Talk: Animal Intelligence and the Evolution of Consciousness*. New York, Simon & Schuster, 1998.
6. Budiansky, S. *Academic roots of paranoia: The Unabomber may not be such an intellectual loner. U.S. News and World Report*, May 13, No. 19. p. 33, 1996.
7. Low, P. The Cambridge *Declaration on Consciousness*. Proceedings of the Francis Crick Memorial Conference, Churchill College, Cambridge University, July 7, 2012, pp. 1–2. Article available at: https://fcmconference.org/
8. Fox, M.W. (ed). *Abnormal Behavior in Animals*. Philadelphia PA. W.B. Saunders Co., 1968.
9. de Wall, B.M. Frans, and Andrews, K. The question of animal emotions. *Science*, 2022; 375: 1351–1352. https://doi.org/10.1126/science.abo2378
10. Marzluff, J.M. Bridging the empathy gap for invertebrates. *Animal Sentience*, 2020; 29(22): 1–3. https://doi.org/10.51291/2377-7478.1614
11. *Veterinary Record*, Editorial. 10 December 2020–2 January 2021. p. 468.
12. CNN. *Sentient Beings: Crabs, octopus and lobsters*. November 22, 2021. Article available at: https://www.cnn.com/2021/11/22/europe/uk-sentient-beings-crabs-octopus-and-lobsters-scn/index.html
13. Fox, M.W. *Bringing Life to Ethics: Global Bioethics for a Humane Society*. Albany NY, State University of New York Press, 2001.
14. Schweitzer, A. *Reverence for Life*. New York, Harper & Row, 1969.
15. Anderson, D.J. *The Nature of the Beast: How Emotions Guide Us*. New York, Basic Books, 2010.
16. Gorvett, Z. *Why insects are more sensitive than they seem*. November 26, 2021. Article available at: https://www.bbc.com/future/article/20211126-why-insects-are-more-sensitive-than-they-seem
17. Grandin, T., and Johnson, C. *Animals in Translation*. New York, Scribner. 2005.
18. Dona, H.S.G. et al. Do bumble bees play? *Animal Behaviour*, 2022; 194: 239–251. https://doi.org/10.1016/j.anbehav.2022.08.013

19. Alumets J. et al. Neuronal localisation of immunoreactive enkephalin and beta-endorphin in the earthworm. *Nature*, 1979 Jun 28; 279(5716): 805–806.
20. Sneddon L.U. Evolution of nociception and pain: Evidence from fish models. *Philosophical Transactions of the Royal Society of London. Series B, Biological Sciences*, 2019 Nov 11; 374(1785): 20190290.
21. Bernhardt B.C., and Singer T. The neural basis of empathy. *Annual Review of Neuroscience*, 2012; 35: 1–23.
22. Panskepp, J., and Panskepp, J.B. Toward a cross-species understanding of empathy. *Trends in Neuroscience*, 2013 Aug; 36(8): 489–496.
23. Griffin, D.R. *The Question of Animal Awareness*. New York, Rockefeller University Press, 1976.
24. Grimm, D. What Are Farm Animals Thinking? *Science*, 2023. https://doi.org/10.1126/science.z2pvw34
25. Yang, T. et al. Hypothalamic neurons that mirror aggression. *Cell*. 2023; 186: 1195–1211. https://doi.org/10.1016/j.cell.2023.01.022
26. Acharya, S.S. and Shukla, S. Mirror neurons: Enigma of the metaphysical modular brain. *Journal of Natural Science, Biology, and Medicine*, 2012 Jul–Dec; 3(2): 118–124.
27. Marshall, J. Mirror neurons. *PNAS*, May 2014; 111: 6531. https://doi.org/10.1073/pnas.1404652111
28. Balcombe, J. *Super Fly: The Unexpected Lives of the World's Most Successful Insects*. New York, Penguin Books, 2021.
29. Fox, M.W. *The Behavior of Wolves, Dogs and Related Canids*. London, Jonathan Cape, 1971.
30. Ryder, R.D. *Animal Revolution: Changing Attitudes towards Speciesism*. London. Blackwell, 1989.
31. Descartes, R. [1637]. *A Discourse on Method: Meditations and Principles*. Translated by Veitch, J. London, Orion Publishing Group. 2004.
32. Darwin, C. *The Expression of the Emotions in Man and Animals*. London, John Murray, 1872.
33. Alderton, D. *Animal Grief: How Animals Mourn*. Dorchester, England, Hubble and Hattie, 2009.
34. Kaswan, P. and Roy, A. Unearthing calf burials among Asian elephants *Elephas maximus Linnaeus, 1758* (Mammalia: Proboscidea: Elephantidae) in Northern Bengal, India. *Journal of Threatened Taxa*, 2024; 16(2): 24615–24629. https://doi.org/10.11609/jott.8826.16.2.24615-24629
35. Engel, C. *Wild Health: Lessons in Natural Wellness from the Animal Kingdom*. Boston and New York, Houghton Mifflin Co., 2002.
36. Álvaro, M.M. et al. The origins of zoopharmacognosy: How humans learned about self-medication from animals. *International Journal of Applied Research*, 2019; 5(5): 73–79.
37. Bronstein, J.L. (ed). *Mutualism*. Oxford, England, Oxford University Press, 2015.
38. Kropotkin, P. *Mutual Aid: A Factor in Evolution*. Boston, MA, Extending Horizons Books, Porter Sargent Publishers. Reprinted edition 2005.
39. Toomey, D. *Kingdom of Play: What Ball-Bouncing Octopuses, Belly-Flopping Monkeys and Mud-Sliding Elephants Reveal about Life Itself*. New York, Scribners Book Co., 2024.
40. Brooks, H.J.B. and Burghardt, G.M. A review of interspecific social play among nonhuman animals. *Neuroscience & Biobehavioral Reviews*, 2023; 151: 105232. https://doi.org/10.1016/j.neubiorev.2023.105232

41. Meints K. et al. Can dogs reduce stress levels in school children? Effects of dog-assisted interventions on salivary cortisol in children with and without special educational needs using randomized controlled trials. *PLOS ONE*, 2022; 17(6): e0269333.
42. The New York Declaration on Animal Consciousness. April 19, 2024, New York University. Article available at: https://sites.google.com/nyu.edu/nydeclaration/declaration
43. Gibbons, M. et al. Can insects feel pain? A review of the neural and behavioural evidence. *Advances in Insect Physiology*, 2022; 63: 155–229.
44. Barrett, M. New frontiers in animal welfare science: An introduction to (farmed) insects. *Veterinary Record*, 2024; 194: 344–346.
45. Frank, E.T. et al. Wound-dependent leg amputations to combat infections in an ant society. *Current Biology*, 2024; 34(14): 3273–3278.e3. https://doi.org/10.1016/j.cub.2024.06.021
46. Oren, G. et al. Vocal labeling of others by nonhuman primates. *Science*. 2024;385:996–1003.

2 Determining Animals' Quality of Life

Criteria and Assessment

There are ethical inconsistencies in the treatment of animals, which are accepted based on various cultural, religious, and economic ideologies and values rather than on a scientific, biological, and ethological understanding of animal sentience and the conditions and care required to provide them with a life worth living. Such consideration is a moral imperative regardless of custom and pecuniary interests. This calls for a full accounting in terms of costs to the animals and the environment and the risks and benefits to humans and society and our duty to provide the best care within the context of their utility. Quality-of-life (QOL) assessments can provide a sound basis for implementing animal welfare standards and reforms, especially in the treatment of farmed and laboratory animals and those used in entertainment and for draft work. When welfare improvements are not implemented, QOL determinations can provide a legal and ethical impetus for the adoption of non-animal alternatives and for the establishment of more effective animal welfare and protection laws and appropriate amendments to existing codes and statutes.

Culturally accepted and economically rationalized ways of keeping animals, such as sows in gestation crates, laying hens and songbirds in cages, apartment dogs in crates all day, and house cats allowed to roam free, need to be reevaluated from the perspective of animals' QOL; related best interests; and social, public health, and environmental consequences.

The intention here is not to provide an exhaustive review of the literature on this topic, about which there has been considerable coverage (see *JAVMA Archives* on QOL), but to offer an integrated approach and objective criteria to help determine the QOL in animals across all situations and contexts, including domesticated, captive, and wild species and whether they are healthy, ill, or injured. This is not to discredit more subjective and experience-based assessments, which may significantly facilitate objective evaluation and interpretation.

There has been considerable research and discussion over the past several years since the UK's Farm Animal Welfare Council presented basic criteria deemed essential for the welfare of animals that addressed both the physical and mental states of animals (1). Known as the Five Freedoms, they are considered by

DOI: 10.1201/9781003569787-2

the World Organization for Animal Health (previously known as the Office International des Epizooties or OIE) when developing international standards on animal welfare (2). The Five Freedoms are: 1. Freedom from hunger and thirst; 2. freedom from thermal and physical discomfort; 3. freedom from pain, injury, or disease; 4. freedom to express normal behavior; and 5. freedom from fear and distress. Paramount is determining and correcting compromised physiological and/or behavioral homeostasis and related cognitive and affective functions to achieve physical and psychological well-being.

Significant progress has been made since these basic freedoms were established, notably in developing the tools to assess animal pain and in evaluating, refining, and advancing standards of care and determining stress and distress in farmed, laboratory, and companion animals and wild animals in captivity (3). A two-domain (comfort–discomfort and pleasure) model has been proposed to describe QOL in animals (4). More recently a 'Five-Domains Model' to facilitate the grading of *both* negative experience (welfare compromise) and positive experience (welfare enhancement) has been developed as a more effective QOL assessment than the basic 'Five Freedoms' (5).

For animals under certain circumstances, specific QOL criteria tailored to address their clinical condition (e.g., osteoarthritis and chronic pain in dogs, response to cancer chemotherapy, and cardiac disease in cats, as assessed by the animals' owners) can be of significant clinical value in determining the prognosis and effectiveness of treatment (6–11).

QOL assessments are essential in making informed decisions, including selecting appropriate veterinary and behavior-modifying interventions, hospice/palliative care (12), and elective euthanasia when there is evident intractable suffering and a prognosis of non-recovery; seizure and protective custody in cases of cruelty and neglect; and quarantine and special (clinical) needs.

When the clinical condition calls for euthanasia, decision-making can be facilitated with a 0–10 scoring of the basic criteria reflecting the animal patient's physical and emotional state, identified as hurt (pain management), hunger, hydration, hygiene, happiness, mobility, and more good days than bad (13). QOL determinations can also help avoid possible overtreatment at the owner's and patient's expense. Ideally, there should be a team approach and, at a minimum, the attending veterinarian with the animal's caregiver/owner.

QOL determinations toward the end of life in healthy animals being handled, transported, and slaughtered can improve their welfare and be cost saving (14,15).

The qualitative aspects of animals' well-being can be quantified using numerical scores for such indices as intensity, frequency, and duration against a baseline norm for that individual or one of similar age, sex, and species under comparable conditions. The professional opinion will then be based more on objective determination than on subjective impression. Non-invasive measures (e.g., determining cortisol stress levels from fecal rather than blood samples) and using audio-visual monitors to control for the observer effect and presence of the owner/handler may be applicable in many instances.

The following list summarizes the basic criteria for QOL assessment and improvement when indicated and feasible.

Physical Considerations

- Provision of physical safety, hygiene, a shelter appropriate to ambient environment and species- and condition-appropriate optimal ambient temperature, humidity, illumination, ventilation, nutrition, and hydration.

- Adequate living space and surfaces to enable normal physical activities.

- Identify and control disturbing visual, auditory, and olfactory stimuli and use calming sounds, pheromones, etc., where appropriate.

- Assess physical condition; general health; and evidence of disease, parasitism, and injuries.

Psychosocial Considerations

- Freedom from fear and anxiety. Assess the animal's affective (emotional) state, including general demeanor and responsiveness. Provide emotional security as needed.

- Identify and alleviate clinical depression, separation anxiety, and abnormal 'coping' behaviors (e.g., compulsive/anxiety disorders) indicative of stress and distress.

- Assess acute/chronic intensity of stress or pain intensity and duration and the prognosis regarding recovery or relief and appropriateness of the humane decision to euthanize rather than prolong suffering. Can the intensity and duration of *unavoidable* suffering be minimized?

- Evaluate if socioenvironmental conditions meet the species and individual's special needs and expression of natural behaviors, with environmental enrichment for physical and cognitive activities, which many animals seek and enjoy.

- Enable animal's control over the immediate environment (e.g., seek shelter, avoid or engage in social contact) and choice in daily routine, especially for some captive wild species, to help prevent emotional/physical stress and subsequent progression to psychosocial imbalance and clinical disorders (e.g., anxiety and compulsive disorders).

Developmental Considerations

- For immature/sub-adult animals, provision of optimal nurturing and social and environmental conditions necessary to promote healthy behavioral, emotional, and cognitive development.

- What hereditary abnormalities (e.g., extreme brachycephaly and pedomorphism) and congenital disorders that affect quality of life can be rectified?

- What transgenerational epigenetic factors can benefit the health and future QOL of offspring (e.g., improved conditions, nutrition, and handling/socialization of animals before breeding and during gestation)?

Additional Considerations

- Determine the training and ability of caregivers at each level of care, along with their attitude, understanding, expectations and actions, and animals' reactions to them.

- With animals that are difficult and dangerous to handle, prioritize safety, behavior modification, and protected contact rather than reliance on physical and chemical restraint and surgical alteration (declawing, defanging), which can be detrimental to QOL.

- What effects does the animal's QOL have on environmental quality and the QOL of others, and vice versa?

- For draft/work animals, determine the optimal workload and duration, and improve harnessing and cart/plough design to reduce stress and injury.

Advances in ethology and biochemistry enabling more objective determination of animals' stress and distress call for further refinement and elaboration of these basic criteria (16–18). This is especially pertinent considering the recently documented lack of consistency in veterinarians' evaluation of case records concerning animal welfare and possible cruelty and neglect (19,20). Inconsistency may also result when animals, depending on their use, are kept under different husbandry/care conditions, some of which may be acceptable to some evaluators but not to others.

Socially accepted norms of animal care and animal use, along with associated public regard and concern for other animals, have evolved in many cultures in the East and West in terms of regarding companion animals less as possessions or chattel and more as individuals with their own rights and interests and as family members. With growing public concern for endangered species, for the health and welfare of animals in their communities, and of those whose meat and other products they consume, the issue of animals' QOL and professional determination of the same by caregivers, animal scientists, and veterinarians is in demand by protectors and legislators alike. Documentation and quantification of clinical and behavioral signs (21) and physiological and biochemical indices of stress and distress in animals by veterinarians, ethologists, and animal welfare scientists are crucial in assessing and improving the QOL of animals. But in the final analysis, 'Although animal welfare science may enlighten us about different problems and opportunities for the animals, it is not able to give a final verdict on what is best for the animal. To come to a view on this involves an ethical judgement' (22). This is the promising role of veterinary bioethics, where sound science, ethics, and empathy are essential elements in improving animals' QOL under our care (23).

Improving the collective QOL of large numbers of animals raised under confined, intensive production systems may be enhanced by adoption of the 'Three Rs' originally applied to animals used in biomedical research and product testing: Refinement (more humane care and procedures), replacement (using non-animal alternatives), and reduction (24). The ecological, environmental, economic, and public health benefits, in addition to benefits related to fewer food animals being raised for their products, may yet be realized by conscientious consumers supporting more humane and sustainable farming practices – a nascent movement still limited by affordability, availability, and opposition by the livestock community and associated vested interests (25).

QOL concerns can be confounded by personal, religious, and politically contentious right-to-life beliefs. So, the terminally ill and suffering are put on life support and unadoptable dogs and cats are incarcerated for life in 'no kill' shelters, many with chronic health issues. In some shelters, cats are neutered and released into unprotected environments by those who see euthanasia as violating the right to life. To not consider each individual animal's QOL and claim they have a right to life regardless of the consequences is tantamount to abdicating ultimate responsibility as their guardians/caregivers. The consequences of invoking the right to life for others, coupled with the doctrine of *ahimsa* (not harming or killing), which prohibits compassionate mercy killing and death with dignity, results in much otherwise avoidable suffering in some countries such as India and in some segments of society (26). The *ahimsa* doctrine mirrors the human-centered pro-life ideology in the less secular West, of which the 'no kill' animal rescue movement is an extension.

FIGURE 2.1 New York City cat holding facility for a trap–neuter–return initiative, which is questionably effective and inhumane.

Now, with the gradual societal and academic acceptance of animals' emotional needs and mental well-being, advances in research in animal welfare science and cognition continue. The gap between our understanding of animals and their needs, emotional states, and consciousness is closing so that better ways to care for them may be implemented (27). As history informs, the wider this gap, the more inhumanity becomes embedded as a cultural norm.

Animals serve society in a multitude of ways as human companions, with many of them bringing the 'good medicine' of trust, friendship/affection, and joy into our daily lives, and more. In the UK, the duty of caring for animals has been codified into law for pet owners to comply with meeting the 'five welfare needs' of their animals, covering optimal housing, diet, behavior, social interactions, and health (28). Society also recognizes the economic, ecological, and other values of domesticated and wild animals, which it affirms with appropriate animal welfare legislation, protection, and conservation laws and conventions. The central role of the veterinary and other animal care professionals in assessing animals' QOL will continue to be challenged by the status quo and cultural norms of animal care and use and is worthy of this task for the welfare and well-being of all (29) under the banners of One Welfare and One Health.

Several models and principles have been proposed to better establish standards of animal welfare and QOL. These include *Three Ethical Concerns* (freedom, feelings, and function); the *Five Freedoms* (freedom from hunger and thirst; freedom from thermal and physical discomfort; freedom from pain, injury, or disease; freedom to express normal behavior; freedom from fear and distress); the *Five Domains* (nutrition, environment, health, behavior, and mental health); and the *Five Welfare Needs* (suitable environment; diet; ability to express normal behaviors; need to be housed with or apart from other animals; and need to be protected from pain, suffering, injury, and disease). These and other similar models or principles are key aspirational or outcome-led frameworks aimed at safeguarding animals under human custodianship. They are widely used in legislation, guidance documents, and protocols. Warwick et al. (30) investigated the effectiveness of these animal welfare models as intended protections and identify several limitations especially concerning captive 'exotic' pets. They reported that animal welfare concerns and problems were regularly identified in relation to both commercial and home environments across all animal classes: Invertebrates, fish, amphibians, reptiles, birds, and mammals, indicating the need for more animal-centered determinations and standards of QOL. Recent advances in the understanding of animal sentience, cognition, and environmental and socioemotional needs make such animal-centered determinations more feasible. The attitude and training of animal handlers and caregivers are also considerable across all phyla.

HELPING THE DISPIRITED

Along with other veterinarians and ethologists (animal behavioral scientists), I have long advocated social and environmental enrichment for animals who become dispirited under the conditions in which they are kept. I reject the clinical term 'depression' because it is too simplistic, and the common remedy for humans, and increasingly for companion animals, is to prescribe antidepressant drugs that too often have harmful side effects.

Animals in relative social isolation, like monkeys in laboratory cages, dogs caged all day while their owners are away at work, pet parrots in small cages, sows on factory farms in narrow crates, circus and Indian temple elephants chained all the time, and various animals alone in barren zoo cages are examples of conditions that are all too common, where animals become dispirited to varying degrees. In extreme cases they become withdrawn; refuse to eat; and develop abnormal, so-called stereotypic behaviors, obsessive-compulsive repetitive movements, and even self-mutilate or become unpredictably aggressive and dangerous to handle. Their immune and endocrine systems can become compromised, leading to secondary health problems. Where there is no relief or escape, learned helplessness and Stockholm syndrome, as in the forced restraint in 'breaking' wild mustangs and captured Asian elephants, often set in.

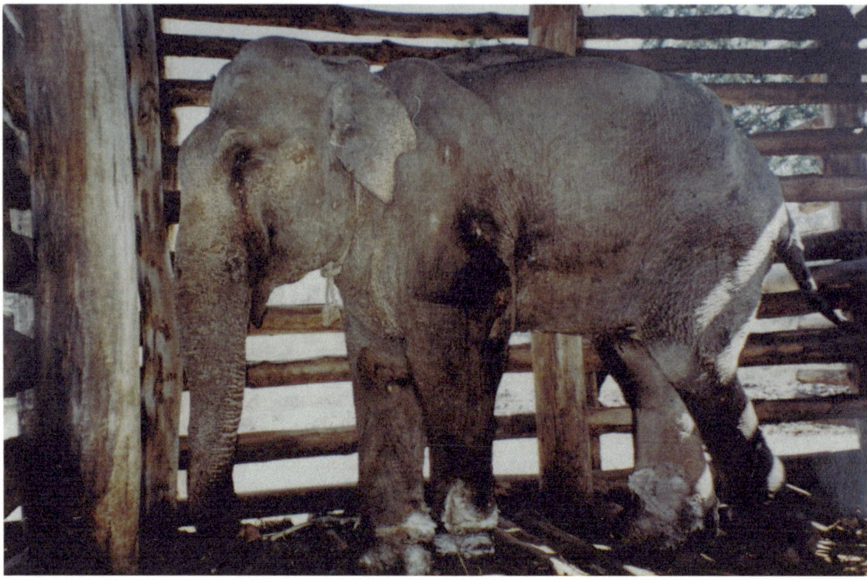

FIGURE 2.2 Captured wild Indian elephant in a corral deprived of proper care and quality of life.

These examples are illustrative not only of our inhumanity and insensitivity but also of some analogous conditions that make us dispirited and which can lead many down the path of drug addiction and suicide. We and other animals share the will to be, the thwarting of which is detrimental to physical and mental health. All sentient beings need life-affirming and enriching experiences with each other and a stimulating environment that engages their senses and physical and mental activities. One of the best and most well-documented medicines is getting out with one's dog and children in nature and spending less time on a computer or smartphone, the detrimental consequences of which are now more clearly understood.

COMPANION ANIMALS AND THE DUTY OF CARE

Many illnesses and behavioral problems in dogs, cats, and other companion animals can be prevented, and others cured, by their caretakers/guardians adhering to five basic principles. These principles combine to make a simple formula to help ensure animals' health and overall well-being:

1. Right understanding and relationship
2. Right breeding/genetics
3. Right nutrition
4. Right environment
5. Right holistic veterinary care

It is every person's responsibility as an animal lover and care provider to recognize the importance of these principles as basic animal rights for several reasons. Following these principles will prevent and alleviate much animal suffering, as well as reduce veterinary and other related costs associated with many animal health and behavioral problems and a reduction in animal abandonment and euthanasia. These principles bring out the best qualities in people as caregivers by enhancing the human–non-human animal bond. Animals receiving quality care will enjoy a higher QOL and relational/emotional experience. The guidelines also provide an ethical compass of responsibility and compassion to advance the moral/character development of children, who, in learning by example how to respect and care for other animals, enhance their self-esteem and self-worth through loving service, and in the process refine their ability to empathize with other sentient beings.

Nearly two dozen countries including England, Australia, and Japan have prohibited or severely restricted the declawing of cats. Exceptions are given to owners who are immunocompromised and at risk from infection from cat scratches rather than euthanizing or rehoming their cats. More than a dozen states have sought to ban the practice, but only New York and Maryland succeeded in 2024. Most encouraging is that America's largest provider of veterinary services, Mars, Inc., with approximately 2,200 animal hospitals, has opted out of declawing, with a position paper stating that removal of the third toe bone from each digit of a cat's foot 'has been shown to lead to chronic pain and, in some cases, to cause long-term behavioral issues.'

Our duty to care for our animals, with or without love, calls for basic knowledge of animal communication and behavior and understanding of animals' needs and for us to have the resources, ability, and commitment to provide for the same. Love alone, without real understanding and basic knowledge of animal care, will not suffice in fulfilling this duty and may cause more harm than good.

While domesticated animals serve us in many ways, those in the wild provide ecological services that are now being recognized as elements of biodiversity and environmental health. QOL in terms of the global life community depends on biodiversity that scientists around the world are now assessing for conservation purposes. This is the complex issue of the next chapter, representing one of the stabilizing factors against climate change and a cornerstone of One Health, calling for multiple correctives that will be proposed.

POSTSCRIPT: 'EXOTIC' (NON-DOMESTICATED) PETS

A useful guide is available enabling one to assess a scale of easy, moderate, difficult, and extremely difficult which exotic species to consider or avoid keeping. See https://www.zoocheck.com/exotic-pets/assessment-enforcement-info/

For more background information, explanation, and worked examples, please refer to the full article:

Warwick, C. et al. (2013). Assigning degrees of ease or difficulty for pet animal maintenance: The EMODE system concept. *Journal of Agricultural and Environmental Ethics*.

REFERENCES

1. Farm Animal Welfare Council (FAWC) (renamed to Animal Welfare Committee (AWC) on 1 October 2019). *Five Freedoms*. 2009. Article available at: http://www.fawc.org.uk/freedoms.htm; for more information on the Five Freedoms visit: https://www.gov.uk/government/groups/farm-animal-welfare-committee-fawc
2. Mullan, S. Assessment of quality of life in veterinary practice: Developing tools for companion animal carers and veterinarians. *Veterinary Medicine*, 2015; 6: 203–210.
3. McMillan, F.D. Quality of life in animals. *Journal of the American Veterinary Medical Association*, 2000; 216: 1904–1910.
4. Mellor, D.J. Updating animal welfare thinking: Moving beyond the 'Five Freedoms' towards 'A Life Worth Living.' *Animals*, 2016; 6: 21.
5. Hielm-Björkman, A.K. et al. Reliability and validity of a visual analogue scale used by owners to measure chronic pain attributable to osteoarthritis in their dogs. *American Journal of Veterinary Research*, 2011; 72: 601–607.
6. Wiseman-Orr, M.L. et al. Validation of a structured questionnaire as an instrument to measure chronic pain in dogs on the basis of effects on health-related quality of life. *American Journal of Veterinary Research*, 2006; 67: 1826–1836.
7. Wojciechowska, J.I. et al. Evaluation of a questionnaire regarding nonphysical aspects of quality of life in sick and healthy dogs. *American Journal of Veterinary Research*, 2005; 66: 1453–1460.
8. Freeman, L.M. et al. Development and evaluation of a questionnaire for assessment of health-related quality of life in cats with cardiac disease. *Journal of the American Veterinary Medical Association*, 2012; 240: 1188–1193.
9. Budke, C.M. et al. Evaluation of a questionnaire for obtaining owner-perceived, weighted quality-of-life assessments for dogs with spinal cord injuries. *Journal of the American Veterinary Medical Association*, 2008; 233: 925–930.
10. Iliopoulou, M.A. et al. Development of a survey instrument to assess health-related quality of life in small animal cancer patients treated with chemotherapy. *Journal of the American Veterinary Medical Association*, 2013; 242: 1679–1687.
11. View AVMA's Guidelines for Veterinary Hospice Care 2016.
12. Villalobos, A. *Quality of Life Scale*. 2008. Article available from: http://pawspice.com/q-of-l-care/new-page.html (Accessed November 1, 2016).
13. Grandin, T. Assessment of stress during handling and transport. *Animal Science*, 1997; 75: 249–257.
14. Grandin, T. Auditing animal welfare at slaughter plants. *Meat Science*, 2010; 86: 56–65.
15. Moberg, G.P. and Mench, J.P. (eds). *The Biology of Animal Stress: Basic Principles and Implications for Animal Welfare*. New York, CABI Publishing, 2000.
16. Möstl, E. and Palme, R. Hormones as indicators of stress. *Domestic Animal Endocrinology*, 2002; 23: 67–74.

17. National Research Council (US) Committee on Recognition and Alleviation of Distress in Laboratory Animals. *Recognition and Alleviation of Distress in Laboratory Animals.* Washington DC, National Academies Press, 2008.
18. Baumgartner, H. et al. Assessment of unnecessary suffering in animals by veterinary experts. *Veterinary Record*, 2016; 179: 307.
19. Morton, D.B. Animal suffering and expert evidence. *Veterinary Record*, 2016; 179: 305–306.
20. Smith, J.L. et al. Discrimination of healthy versus sick steers by means of continuous remote monitoring of animal activity. *American Journal of Veterinary Research*, 2015; 76: 739–744.
21. Sandoe, P. et al. *Companion Animal Ethics.* Ames, IA, Wiley-Blackwell, 2016.
22. Fox M.W. Veterinary bioethics. In Schoen, A.M. and Wynn, S.G. (eds.), *Complementary and Alternative Veterinary Medicine: Principles and Practice* (pp. 673–678). St. Louis, MO, Mosby, 1997.
23. Fox, M.W. *Laboratory Animal Husbandry: Ethology, Welfare and Experimental Variables.* Albany, NY, State University of New York Press, 1986.
24. Fox, M.W. *Eating with Conscience: The Bioethics of Food.* Troutdale, OR, New Sage Press, 1997.
25. Krantz, D.L. and Fox, M.W. *India's Animals: Helping the Sacred and the Suffering.* CreateSpace Books, Amazon.com, 2016.
26. McMillan, F.D. (ed) *Mental Health and Well-Being in Animals.* Ames, IA, Blackwell Publishing, 2005.
27. BVA/RCVS Vet Futures Action Plan. Available from: https://vetfutures.org.uk/resource/vet-futures-action-plan-2016-20/ (Accessed November 19, 2016).
28. Colonius, T.J. and Earley, R.W. One welfare: A call to develop a broader framework of thought and action. *Journal of the American Veterinary Medical Association*, 2013; 242: 309–310.
29. Pinillos, R.G.R. et al. One welfare—a platform for improving human and animal welfare. *Veterinary Record*, 2016; 179: 412–413.
30. Warwick, C. et al. Are the key welfare models effective for exotic pet animals? *Discover Animals*, 2024; 1: 15. https://doi.org/10.1007/s44338-024-00013-2

3 The Importance of Natural Biodiversity for Health and Well-Being

Saving and restoring biodiversity – indigenous plant and animal species and microorganisms – is essential for the health of aquatic and terrestrial ecosystems (which are models of how our industrial economy should function sustainably). This will help secure life-giving and health-sustaining 'ecological services,' including air and water purification, flood and drought control, nutrient cycling, pollination of crops and wild plants, and control of invasive pests and diseases.

Cultural as well as biological diversity of indigenous species and peoples, along with indigenous knowledge, traditional food production and plant-based medicinal practices, crops, and domestic animal varieties, call for greater protection under the banner of One World, One Health, One Economy. As responsible citizens, we must make the climate and extinction crises the top priority for responsible governments and industries to address and call for the establishment of a United Environmental Nations to help save and restore critically compromised human communities and wildlife habitats – aquatic and terrestrial.

It is evident from the United Nations 2019 IPBES Global Assessment Report on Biodiversity and Ecosystem Services that we have yet to learn to share this planet with other intelligent life forms and consciousnesses (1). This would be for our own good, which is bound to that of other beings who comprise and sustain the life community, even if we do not respect and care for the least of them. The report finds that around 1 million animal and plant species are now threatened with extinction, many within decades, more than ever before in human history. The average abundance of native species in most major land-based habitats has fallen by at least 20%, mostly since 1900. More than 40% of amphibian species, almost 33% of reef-forming corals, and more than a third of all marine mammals are threatened. The picture is less clear for insect species, but available evidence supports a tentative estimate of 10% being threatened. At least 680 vertebrate species have been driven to extinction since the 16th century, and more than 9% of all domesticated breeds of mammals used for food and agriculture have become extinct by 2016, with at least 1,000 more breeds still threatened.

DOI: 10.1201/9781003569787-3

Humans are the most invasive of all species. The later UN Biodiversity Conference in Montreal in December 2022 succeeded in securing non-binding agreements from close to 200 countries to protect 30% of terrestrial and aquatic ecosystems by 2030. The main environmentally destructive drivers identified that are reducing biodiversity and accelerating species extinction were agriculture, overfishing, logging, mining, climate change, invasive species, and pollution. But there were no media reports to my knowledge of any substantial discussion about the issue of population growth, with our numbers now being estimated at 8 billion.

Many UN delegates and conferees were not even born when this issue was first raised by the Club of Rome's report, *The Limits to Growth*, first published in 1972 (2). Widely publicized to world leaders and governments (and subsequently updated), the report showed that economic growth could not continue indefinitely because of resource depletion. It underscored as self-evident that economic growth meant greater environmental impacts, and the main way to environmental gains was reduced population and consumption.

The natural diversity of animals, including insects, plants, and soil and water microorganisms that are native to a particular place (ecosystem), helps maintain environmental health and sustainability as well as the food web. Such communities also help prevent the spread of invasive and potentially harmful species.

These controls break down and ecological dysbiosis sets in because of various human activities, especially industrial farming, aquaculture, and agroforestry, with virtual monocultures that drastically reduce biodiversity. Invasive weeds, pests, and diseases take hold. Rather than change practices and adopt organic, more ecologically sound systems of production, the treadmill of reliance on herbicides, fungicides, insecticides, antibiotics, antiparasitics, and other drugs has become the accepted, profit-driven norm.

Public sanitation, personal hygiene, humane animal husbandry, and environmental stewardship all help avoid the need for the 'preventive' use of such chemical agents and their harmful consequences. Like antibiotics and fungicides, insecticides kill many non-target, harmless species that help control the proliferation of other potentially harmful species through competitive exclusion. The public is also part of this collateral damage, along with honeybees and other pollinators.

The loss of biodiversity and its trophic, regulatory effects coupled with drug and chemical misuse and herbicide-resistant, insecticide-producing corn and other genetically modified (GM) crops all help create 'superbugs' – resistant strains of bacteria, fungi, and other pests and diseases. Some are now resident in many hospitals; herbicide-resistant 'superweeds,' blights, and pests constantly threaten and ruin crops; drug-resistant populations of mosquitoes and internal parasites flourish around the world. Declining mosquito-consuming frog and bat populations mean a rise in the incidence of malaria and other mosquito-borne diseases.

FIGURE 3.1 Insect–plant relationships are integral aspects of sustainable biodiversity.

The rising incidence of Lyme and other tick-borne diseases in the Americas is not simply attributable to climate change but many other factors, especially wildlife mismanagement encouraging high deer populations (for the hunting industry) and extermination of wolves and cougars (for the livestock industry), along with a paucity of foxes and other eaters of small rodents that are disease reservoirs and of insectivorous birds and reptiles that consume disease-vectoring ticks.

Additional health and environmental problems are created by the importation of plant and animal produce and animal feed from abroad, which calls for heightened biosecurity, along with legislation, to prohibit the ownership, propagation, and deliberate or accidental release of 'exotic' animals and plant varieties. In a mere 3 years, for example, the fire salamander has declined by 99.9% in Holland due to the chytrid fungus disease originating in Asia from imported amphibians. Also, domestic animals that become feral, from cats and camels to cattle and pigs, are a major threat to many wild species and biodiversity on many continents. Animal health and well-being must be integrated with human and environmental well-being.

Optimal biodiversity is the keystone of One Health – animal, plant, environmental, and human – which Albert Schweitzer MD, was one of the first to recognize in his call for reverence for all life, an ethic that contradicts the ethos of consumptive and destructive societies. Having respect for all beings and things is one of the seven traditional core values of the Lakota Sioux. Spiritually, the One Health concept is the acknowledgement of our sacred connections expressed as respect for all life and care for the well-being of all living things, not just human beings. The 'medicine wheel' is the Anishinaabe (Native American Indian) visionary and ethical equivalent of One Health, which also incorporates right relationships and right action.

Now we face what scientists and conservationists call the sixth mass extinction, the Anthropocene apocalypse, the severity of which we may be able to minimize given the will and the means. Climate change and loss of biodiversity go together and will be the 'natural' population and consumption correctives, along with associated famines, plagues, and pestilences, if we fail to act now.

The U.S. government and the Vatican declined to sign the agreement at the 2022 UN's COP15 biodiversity summit in Montreal. COP15's goal was to stop the global loss of biodiversity by 2030, and representatives from Cameroon, the Democratic Republic of Congo, and Uganda expressed disagreement. Signatories agreed to take steps to halt species extinction, protect 30% of the planet for nature, restore 30% of degraded ecosystems, and reform $500 billion in subsidies that encourage environmental damage (3).

One detailed review of these concerns states:

- 'Biodiversity experts estimated that about 30% (uncertainty range: 16–50%) of species have been globally threatened or driven to extinction since the year 1500.
- There was overwhelming consensus that global biodiversity loss will likely decrease ecosystem functioning and nature's contributions to people.
- Global biodiversity loss and its impacts may be greater than previously thought, due to higher estimates provided for understudied taxa and by underrepresented experts.
- Experts estimated that greatly increasing conservation investments and efforts now could remove the threat of extinction for one in three species that may otherwise be threatened or extinct by the year 2100. A recent estimate that at least one million species of animals and plants are currently threatened with extinction assumed that 10% of insect species are threatened, based on a comprehensive review of the limited available evidence. Our survey estimates, which were provided by 629

> experts who study terrestrial and freshwater invertebrates, therefore suggest that the percentage of insect species that are threatened may be much higher' (4).

Electromagnetic radiation is an emerging driving factor for the decline of insects, according to entomologist Alfonso Balmori, who writes, 'The biodiversity of insects is threatened worldwide. Numerous studies have reported the serious decline in insects that has occurred in recent decades. The same is happening with the important group of pollinators, with an essential utility for pollination of crops. Loss of insect diversity and abundance is expected to provoke cascading effects on food webs and ecosystem services' (5). Many authors point out that reductions in insect abundance must be attributed mainly to agricultural practices and pesticide use. On the other hand, evidence for the effects of non-thermal microwave radiation on insects has been known for at least 50 years. The review carried out in this study shows that electromagnetic radiation should be considered seriously as a complementary driver for the dramatic decline in insects, acting in synergy with agricultural intensification, pesticides, invasive species, and climate change. The extent that anthropogenic electromagnetic radiation represents a significant threat to insect pollinators is unresolved and plausible. For these reasons and considering the benefits they provide to nature and humankind, the precautionary principle should be applied (5).

CARING FOR WILDLIFE: REDUCING FERAL ANIMAL INSURGENCE

Caring for all creatures great and small who are still surviving in the diminishing wild by ensuring their well-being is a responsibility all communities and countries must adopt. The harmful incursions of people, free-roaming and feral cats, dogs and pigs, and unattended cattle cannot continue to be ignored.

Domestic animals that adapt to living in the wild are called feral. Feral pigs are becoming an increasing problem in the United States and Canada (6). Feral pigs put domestic pigs at risk from diseases such as African swine fever. Native to Eurasia, these animals were first introduced to Canada in the 1980s as exotic livestock for meat. Resulting from farm escapes and intentional releases, wild pigs have already caused widespread problems in Canada's prairie provinces and many American states. Pseudorabies was found in samples taken from two feral pigs captured near Burson in Calaveras County, California, in January 2024. The virus is not related to rabies but can kill dogs and cats and can cause reproductive problems in livestock, including pigs, cows, sheep, and goats.

There is also a significant feral dog problem (7). Feral dogs have been documented in all 50 states, and estimates of damage in the United States from these animals amount to >$620 million annually. In Texas alone, it is estimated that over $5 million in damage to livestock annually can be attributed to feral dogs (8).

Feral and owned free-roaming dogs spread diseases such as distemper and parvovirus that put endangered and protected species like the wolf at risk. According to one analysis,

> The domestic dog (*Canis familiaris*) is found on every continent that humans have settled. More than 70% of the global dog population (estimated at >700 million to ~1 billion) comprises of free-ranging dogs (FRD). In many developing countries, FRD are associated with the transmission of zoonotic diseases such as rabies, zoonotic visceral leishmaniasis, canine echinococcosis, and soil borne helminths. Rabies alone is responsible for an estimated 60,000 human deaths per year worldwide, with most deaths occurring in Asia and Africa. In addition, FRD are an important and emerging threat for livestock as well as biodiversity. Furthermore, FRD also suffer from poor health, high mortality, and abuse. Amongst the countries with the largest FRD populations in the world, India stands out as it accounts for an estimated 20 million dog bite cases per year and around 20,000 dog-mediated human rabies deaths per annum. Apart from being a hotspot of dog-mediated rabies deaths, dog attacks also result in direct human fatalities in India. There is thus a strong and urgent need to control free-roaming dog populations. In general, efforts to control dog populations in India using a variety of lethal and non-lethal methods have been unsuccessful so far, even though this has been attempted for over 200 years. (9)

Domestic dogs have contributed to 11 vertebrate extinctions and are a known or potential threat to at least 188 threatened species worldwide (10). They also compete with wild predators and scavengers for food.

Both owned and unvaccinated and feral dogs and cats pose a greater public health risk, as potential carriers of rabies, than do wild carnivores such as coyotes and foxes. The feral cat problem is also a major factor in the loss of songbirds and small mammals and the spread of diseases to wild felids. In the extensive review on this issue, the authors state, 'Domestic cats have also been implicated at broader scales, in the global extinction of at least 63 species – 40 birds, 21 mammals, two reptiles – which is to say 26% of all known contemporary extinctions in these species groups. Likewise, domestic cats currently endanger at least a further 367 species which are at risk of extinction' (11).

All told, an estimated 2,036 species fall victim to free-roaming domestic cats. This adds up to 9% of all known birds, 6% of mammals, and 4% of reptiles. Cats have been documented eating creatures on every continent, even Antarctica. In a global assessment of the impact of feral and owned, free-roaming domestic cats, the authors state:

> Free-ranging cats (*Felis catus*) are globally distributed invasive carnivores that markedly impact biodiversity. Here, to evaluate the potential threat of cats, we develop a comprehensive global assessment of species consumed by cats. We identify 2,084 species eaten by cats, of which 347 (16.65%) are of conservation concern. Islands contain threefold more species of conservation concern eaten by cats than

FIGURE 3.2 Free-ranging cats are globally distributed invasive carnivores that markedly impact biodiversity.

> continents do. Birds, reptiles, and mammals constitute ~90% of species consumed, with insects and amphibians being less frequent. Approximately 9% of known birds, 6% of known mammals, and 4% of known reptile species are identified in cat diets. 97% of species consumed are <5 kg in adult body mass, though much larger species are also eaten. The species accumulation curves are not asymptotic, indicating that our estimates are conservative. Our results demonstrate that cats are extreme generalist predators, critical for understanding their impact on ecological systems and developing management solutions. (12)

Keeping cats indoors and providing regular veterinary wellness examinations are the best ways to help prevent the spread of toxoplasmosis, a disease caused by the parasite *Toxoplasma gondii*. The domestic cat is the definitive host that harbors and spreads this parasite to humans and other species when allowed to roam free. In a groundbreaking study, a group of international scientists reported on the potential contribution to the signs of frailty in older adults, including exhaustion and muscle loss, to infection with this parasite (13). Another review on the effects of this parasite on humans concludes that toxoplasmosis could play a substantial role in the etiopathogenesis of mental health disorders, and its association with schizophrenia is the second strongest association, after autism (14). Other studies have correlated infection with this parasite contributing to impulsive acts of violence, which forensic psychiatrists and criminologists should consider.

These feral animal issues are occurring worldwide and include America's 'wild' mustangs – domestic horses in the West that have become feral – cattle and camels in Australia, as well as rabbits and cats, where it is estimated that invasive species are beginning to outnumber indigenous ones, many of which they kill or outcompete. In February 2023, 65 cattle, deemed feral, were shot from the air in the Gila Wilderness area in the state of New Mexico according to the Forest Service. The shooters were agents with the Department of Agriculture's Animal and Plant Health Inspection Service (APHIS). Some APHIS agents specialize in killing predators and nuisance animals. Feral cattle are also an issue in Hawaii and several other states, including California and Utah.

Collectively, the negative impact of feral animals on biodiversity must be addressed. The World Economic Forum lists biodiversity loss as one of the top three global risks, because the loss of nature does not just impact plants and animals. Destruction of forests, wetlands, and other natural ecosystems deprives us increasingly of clean air and water and results in the release of 'emergent' diseases, pestilence, and pests, which natural biodiversity helps contain.

We must make fundamental changes in how we treat domesticated animals for food, as pets, and for other purposes and those who become feral at home and abroad. But where there are no or few large indigenous wild herbivores, the ecological benefits of some feral domestic grazers should be considered and their presence accepted as a transitional stage in future rewilding, as proposed in one extensive review of this issue of routinely killing feral 'invasive' species.

> Large mammalian herbivores (megafauna) have experienced extinctions and declines since prehistory. Introduced megafauna have partly counteracted these losses, yet are thought to have unusually negative effects on plants, compared with native megafauna. Using a meta-analysis of 3,995 plot-scale plant abundance and diversity responses from 221 studies, we found no evidence that megafauna impacts were shaped by nativeness, 'invasiveness,' 'feralness,' coevolutionary history, or functional and phylogenetic novelty. Nor was there evidence that introduced megafauna facilitate introduced plants more than native megafauna. Instead, we found strong evidence that functional traits shaped megafauna impacts, with larger-bodied and bulk-feeding megafauna promoting plant diversity. Our work suggests that trait-based ecology provides better insight into interactions between megafauna and plants than do concepts of nativeness. (15)

HUMAN INCURSIONS

The multiple-use concept allowing noisy and polluting all-terrain vehicles, dirt bikes, and snowmobiles into wildlife-inhabited areas may be 'democratic.' The appeals by hikers, naturalists, wildlife photographers, and other lovers of nature to leave these places undisturbed and undefiled and to prohibit such vehicular traffic are generally ignored by state wildlife 'management' authorities.

But an 'ecocratic,' non-anthropocentric stewardship of lands belonging to wildlife would be just and enlightened self-interest because our ultimate well-being is dependent on restoring and protecting natural biodiversity.

Limiting the numbers of visitors to wildlife habitats as well as times and seasons will help reduce wildlife disturbances, disruptions in feeding, breeding, care of young, and associated stressors that increase susceptibility to disease. Core areas should be strictly off-limits and, at the most, trail cameras permitted for non-invasive wildlife research and monitoring. Human hikers create a 'landscape of fear' that chases other animals into hiding, a new study has found. Even when hikers are unarmed and using the landscape peacefully, they can cause disruption on par with that of apex predators (16).

The United States became a world leader in conservation efforts when President Richard Nixon signed the Endangered Species Act (ESA) in 1973 that followed the Clean Water Act, the Clean Air Act, the Wilderness Act, and the National Wild and Scenic Rivers Act. These acts helped empower non-profit conservation, animal, and environmental protection organizations and their civil society influence. These acts also triggered opposition and litigation by various corporate interests and organizations that felt economically threatened. But there was a glimmer of hope in the August 14, 2023, ruling by Montana District Court Judge Kathy Seeley in favor of young environmental activists who claimed state agencies were violating their constitutional right to a clean and healthful environment by allowing fossil fuel development.

OUR FATE AND THE FATE OF THE FORESTS

Our forests are in peril by all accounts, from fires, high winds, and droughts. An estimated 11–16% of U.S. tree species are threatened with extinction, with the most common threat being invasive and problematic pests and diseases. We need the trees to hold the hills and mountainsides to prevent erosion, to sequester carbon emissions, to reduce global warming and climate change, and to produce the oxygen we and most other life forms need to live. But still they are seen primarily as a resource to be harvested and to be cleared for agricultural expansion, now decimating the Amazon rainforest. The ultimate irony in many Asian countries is that spirit-broken elephants are enslaved by the timber industry to drag out the last of the great trees from their jungles and forests from their domain. This is their natural habitat, and their sovereignty, aside from their ecological services, like that of whales and wolves, elephants, and eagles, needs to be recognized as being on par with the sovereignty of nation-states.

What is utterly absurd is that U.S. taxpayers' subsidies for the federal logging program ranged from $1.6 to $1.8 billion per year in fiscal years 2013–2017. 'The Forest Service attempts to justify these losses by hiding commercial timber sale projects within larger ecological restoration projects – a move that consistently lands the agency in court. Selling timber from federal lands below cost is a form of environmentally harmful subsidy that runs afoul of international agreements' (17).

Part of the problem for the trees is invasion by insect pests that have a free range thanks to the spraying of insecticides to kill mosquitoes and the wholesale application of pesticides by the agricultural and agroforestry industries and, in the process, beneficial pollinator insects and pest-controlling parasitic wasps and insectivorous reptiles, amphibians, and birds are either poisoned or starve to death. After all, an estimated 40% of all insect species are declining globally and a third of them are endangered (18).

Like the Dust Bowl of the 1930s across millions of acres of the Great Plains, the fate of America's trees and those of other nations may indeed culminate in the ancient prophetic warning that when the trees are gone, the sky will fall. The famous statement by the late President Franklin D. Roosevelt, 'A nation that destroys its soils destroys itself. Forests are the lungs of our land, purifying the air and giving fresh strength to our people,' is more relevant today than ever. A significant initiative would be for the U.S. government, if it is serious about climate change, to end the timber industry subsidies, which should be diverted to reforestation and conservation. Because of climate change droughts, mega-forest fires that cannot be extinguished are more frequent, exacerbating air pollution with microparticles that are a public health risk.

Property owners can plant indigenous trees and engage in 'rewilding' instead of wasting water, wasting energy mowing, and applying harmful chemicals to lawns and golf courses. As Chief Seattle advised, 'Whatever befalls the earth befalls the sons of the earth. Man did not weave the web of life; he is merely a strand in it. Whatever he does to the web, he does to himself' (19).

Every law student should have to read Christopher D. Stone's book, *Should Trees Have Standing*? (20). We might then eventually have a Supreme Court that better serves the common good and the good of the commons.

I wish my species would acquire the wisdom and the means to live more like trees, giving more to life than it takes.

GRASSLANDS NEED RESTORATION

While Americans decry the climate consequences of Amazon rainforest destruction (which ecologists report has reached a grave state of regenerative impairment), they should take the log out of their own eyes and look at the legacy of the buffalo exterminators and 'sod busters' who continue to destroy the magnificent prairies and high plains, once home to vast herds of grassland-nutrient-cycling buffalo and antelope and the now endangered swift fox and other mammals and birds (21). All states must support the Big River Connectivity Rewilding (22).

My friend, the late Wes Jackson's, dedicated research and development of a perennial wheat he named kernza offers a valuable food and fodder crop to help with prairie soil and habitat restoration and carbon sequestration (23).

The rewilding of grasslands would help restore the depleted and chemically contaminated Oglala aquifer. A reduction in agricultural fertilizer and pesticide use, also given to livestock and poultry along with antibiotics and other drugs, would help reduce contamination of the Great Lakes, where fish now pose a health risk to consumers, especially indigenous Anishinaabe communities. After the genocide of prairie and plains Indian tribes who lived sustainably with the buffalo and other wildlife, it is deplorable that the U.S. government subsidizes cattle ranchers, who turned millions of acres of grasslands into semi-desert public 'rangelands' from grazing too many livestock, and funds predator extermination programs, further decimating biodiversity.

FIGURE 3.3 Timber industry deforestation in Washington State causing erosion and silting of a salmon river habitat.

Tearing up grasslands and using arable land today to produce crops for biofuels is not a 'green' energy source, since production releases greenhouse gases such as carbon dioxide, nitrous oxide, and methane. Likewise, the collection of peat from wetlands and draining of wetlands for crop production on peat soils (histosols) turns a major carbon sink into a methane and carbon dioxide emitter. Significant progress has been made, but more is needed to protect all wetlands so they can sustain natural biodiversity and help control flooding and droughts (24).

The 'rewilding' movement to restore natural terrestrial and aquatic habitats is an essential part of the One Health, One Earth, and One Economy paradigm connecting the security and well-being of our own species with that of others that help sustain our collective planetary life community.

HELPING OUR DECLINING BIRD POPULATION

'One in eight bird species is threatened with extinction, and the status of the world's birds continues to deteriorate: species are moving ever faster towards extinction. For those not yet considered threatened, the majority are in decline and have much-depleted populations. For example, 2.9 billion individual birds are estimated to have been lost in North America since 1970' (25).

FIGURE 3.4 Owned and feral cats are a major contributor to loss of biodiversity as predators of birds and small mammals.

Scientists have identified that habitat loss is the biggest overall driver of bird declines. Habitat loss occurs when land is converted for agriculture, development, resource extraction, and other uses. Habitat degradation is a second cause of losses. In this case, habitat does not disappear outright but becomes less able to support birds, such as when the habitat is fragmented, altered by invasive plants, or water quality is compromised. Compared to wild predators of birds, free-roaming domestic cats are the top contributors to this decline.

The first-ever comprehensive assessment of net population changes in the United States and Canada reveals across-the-board declines that scientists call 'staggering.' All told, the North American bird population is down by 2.9 billion breeding adults, with devastating losses among birds in every biome. Forests alone have lost 1 billion birds. Grassland bird populations collectively have declined by 53%, or another 720 million birds (26). The famed Cornell University Ornithology Department has posted *Seven Simple Ways to Help Birds*:

1. Make Windows Safer, Day and Night. Simple adjustments to your windows can save birds' lives.
2. Keep Cats Indoors. Indoor cats live longer, healthier lives. Outdoor cats kill more birds than any other non-native threat.
3. Reduce Lawn by Planting Native Species. The U.S. has 63 million acres of lawn. That's a huge potential for supporting wildlife.
4. Avoid Pesticides. Look for organic food choices and cut out some of the 1 billion pounds of pesticides used in the U.S. each year.
5. Drink Coffee That's Good for Birds. Shade-grown coffees are delicious, economically beneficial to farmers, and help more than 42 species of North American songbirds.
6. Protect Our Planet from Plastics. 91% of plastics are not recycled, and they take 400 years to degrade.
7. Watch Birds, Share What You See. Bird watchers are one of science's most vital sources of data on how the ecological world is faring.

A solution that is good for cats and birds: Save birds and keep cats healthy by keeping cats indoors or creating an outdoor cat enclosure, or 'catio.' Some cats can be trained to enjoy walks in a harness attached to a leash.

Millions of people travel by air daily for business and pleasure in total disregard for their massive carbon footprint, an activity contributing not only to climate change but also to bird deaths. According to one article on this issue, (27) the Federal Aviation Administration (FAA) says more than 9,000 birds are reportedly struck annually by planes in the United States, a figure that is rising every year. But because pilots do not have to report inconsequential bird strikes, the actual number is likely twice that. The FAA says the rate over the past 23 years has been one a day and that bird strikes cost the airline industry more than $700 million every year.

FIGURE 3.5 The United States has 63 million acres of lawn. That's a huge potential for supporting wildlife like this goldfinch.

THE HIDDEN COSTS OF SEAFOOD

China's distant-water fishing operation is the largest in the world, catching more than 2 billion kilograms of seafood each year – one-fifth of international trade. For this powerful interactive feature, reporters threw messages in bottles to captive ship workers, tracked clandestine fishing operations off the shores of South America, and followed private investigators through China to reveal the intrinsic relationship between overfishing and human rights abuses. This investigation tracked the seafood all the way back to U.S. supermarket shelves, dinner tables, and university cafeterias (28).

Andrew Sharpless, chief executive of Oceana, an international organization focused on ocean conservation, writes that a 'well-managed and abundant ocean could feed a billion people a healthful seafood meal every day, forever. Overfishing, especially by big industrial fleets, is destroying that abundance – collapsing a wild food resource essential to the health and livelihoods of hundreds of millions of people along coastlines around the world' (29).

Now, some major food companies are putting cell-cultivated meat and fish products on the market in line with the consumer trend favoring cultured meats and other animal products (30). Production costs in terms of providing safe cell nutrients and affordability, depending on the scale of production, may complement vegan diets to eliminate animal suffering and killing, as well as the environmental consequences of live animal farming.

Although farmed fish were recognized as sentient under the UK Animal Sentience Act in 2022, they are afforded no legal protection for their welfare at the time of slaughter. Nor are they in the United States. When harvested, they suffocate to death when out of the water. Implementing humane electrical stunning to prevent pain and distress is now being initiated and standards set in the UK, and I hope all other countries involved in fish farming will follow suit.

One-quarter of freshwater fish are at risk of extinction, according to an update to the Red List of Threatened Species maintained by the International Union for Conservation of Nature (IUCN). Data on 1,640 newly evaluated freshwater fish species – combined with the 14,898 already assessed – show that pollution is the greatest threat to the group. And climate change is exacerbating the problem. 'Climate change really compounds all of the other threats,' says Catherine Sayer, who leads the freshwater biodiversity program at the IUCN. Despite the grim update, scientists say that good, collaborative conservation efforts have been successful in changing the fates of several species.

The fate of the oceans, forever contaminated with plastics, as well as the fate of the plankton that release more atmospheric oxygen than all the forests and rising acidification and temperatures that feed extreme weather events, call for united rectification and amelioration.

'GLOBALIZATION' OF THE ECONOMY AND DISEASE

The globalization of the economy, especially through world trade in seedstock, livestock, and food commodities and world travel, also means the global spread of diseases affecting us as well as indigenous plants and animals. This calls for a United Environmental Nations that addresses from a multidisciplinary perspective the best ways to prevent diseases from an integrated One Health perspective. This is the key to economic and public health security and protection of biodiversity.

We can take one lesson from nature in the spread of West Nile virus from birds migrating into the United States that has decimated many indigenous species of birds and infected thousands of people bitten by infected mosquitoes. According to one study, since West Nile virus emerged in the U.S. in 1999, 22,999 neuroinvasive disease cases in humans were reported through 2017. These cases have arisen from an estimated 7 million human infections. Since many infected people are asymptomatic, considerable expense has been needed to screen for the virus in blood donors and for organ transplants to prevent transfer of the virus to recipients (31).

The U.S. government and all responsible authorities should also consider the fact that millions of mammals, amphibians, birds, insects, and reptiles are imported legally into the United States every year, potentially bringing with them 'a kaleidoscope of pathogens,' writes former Fish and Wildlife Service inspector Jonathan Kolby. 'With few exceptions, the US has no laws specifically requiring disease surveillance for wildlife entering the country, and the vast majority of wild animal imports are therefore not tested,' Kolby writes (32).

Many correctives documented by regenerative, carbon-sequestering organic farming practices, permaculture and polyculture that maximize the benefits of optimal biodiversity, and holistic veterinary and human medicine have been marginalized by the mainstream agribusiness food and health-care systems and their allied petrochemical, pharmaceutical, and biotechnology industries. This consumer-driven monoculture will be perpetually at war with invasive species and diseases, and we will suffer the consequences until the principles of democracy are applied ethically, ecologically, and economically for the common good and the good of the commons.

PREVENTING REVERSE ZOONOSES: PEOPLE INFECTING ANIMALS

Animals catching diseases from humans, called reverse zoonoses, affects species around the world. These include mussels contaminated with hepatitis A virus, cheetahs that contract influenza A, tuberculosis transmitted to Asian elephants, and the parasite *Giardia duodenalis* passed on to African painted dogs (*Lycaon pictus*) (33). Because of their evolutionary closeness to humans, great apes are especially vulnerable to our diseases. In a 2018 literature review, researchers documented 33 probable or confirmed disease transmission events from humans to chimpanzees and mountain gorillas. Those events included deadly cases of measles, human metapneumovirus, and the bacterium *Streptococcus pneumoniae* (34). Greater biosecurity is called for to prevent reverse zoonoses by people infected with potential pathogens, especially tourists and indigenous peoples entering wildlife habitats, potentially causing more harm than poaching and agricultural encroachment. Human respiratory viruses and bacteria pose dire threats to endangered great apes, and researchers have found that guidelines to protect great apes are routinely violated, with tourists coughing, sneezing, and even urinating within a few feet of wild apes and touching trees and plants, creating fomites. A recent study also found that adults who have sick children but are themselves asymptomatic may transmit viruses to great apes when they go into forests and wildlife preserves, and researchers are seeking creative ways to improve the health of apes and the people living near them.

Past human civilizations had their plagues and pestilences. These will continue until we identify and prevent the causes of such harmful natural forces and become more civilized in our relationships with other species and the natural environment. Plagues and pestilences arise in cycles in relation to climatic and other environmental changes, local and global. High-density human and domestic animal and plant aggregations that disrupt ecosystem biodiversity are especially vulnerable.

The first correctives and preventives – reduction in population and consumption – would be better taken by ourselves, voluntarily, and with less suffering than is inevitable with nature's correctives of pestilence, plagues, droughts, floods, and famine.

The rewilding and restoration of biodiversity and healthy ecosystems by reintroducing keystone species can help create resilience to climate change and lead to enlightened self-interest in terms of public health and the global economy.

The California Department of Fish and Wildlife and Maidu Summit Consortium have released seven beavers into waters near the tribal community of Tasmam Koyom in hopes of developing a breeding population that will serve as a nature-based ecosystem restoration tool, supporting groundwater recharge and other essential functions. Beavers are a 'keystone species,' and '[t]here's a cascade of effects where you see (once beavers are gone), other species are no longer there,' said department official Valerie Cook (35).

Beavers were almost trapped to extinction by the fur trade, causing ecological damage to millions of acres of wildlands and loss of biodiversity and rainwater containment. On a related issue of wildlife exploitation, the bearskin regimental headgear of the British Royal King's Guard should be buried in shame. Bearskin hats are made from the skin of black bears who contribute in many ways to the health of the forests, hundreds of which are killed annually in Canada as well as in the United States.

Sea otters help control erosion by eating striped shore crabs that burrow into salt marsh banks and eat vegetation, according to a study in *Nature* (36) that found the slowest rates of erosion in California marshes are in areas with the highest populations of sea otters. Sea otters were hunted nearly to extinction, but their numbers have rebounded thanks to hunting bans, habitat restoration, and a program at Monterey Bay Aquarium to raise and release orphaned sea otters.

It is a positive sign of progress that healthy wolves captured humanely in Oregon were translocated to Colorado in the winter of 2023, where wolves have been exterminated, to help restore the natural habitat and control elk and mule deer numbers and thus protect the forest. Also, good news is that some cattle ranchers are adopting rotational grazing, reducing stock numbers, and adopting non-lethal methods of predator control.

The Americas are famed for their cattle-herding and -driving cowboy and gaucho culture, which is mythologized by linking beef with masculinity and virulence.

This next chapter is a 'beef with beef,' the cattle culture, and its barons and will be followed by a chapter on the wolf, whom the cattle industry has persecuted for centuries and all but exterminated across North America.

REFERENCES

1. United Nations 2019 IPBES Global Assessment Report on Biodiversity and Ecosystem Services. https://www.ipbes.net/assessment-reports/eca
2. Meadows, D. et al. *Limits to Growth: The 30-Year Update*. Vermont, Chelsea Green Publishing, 2005.
3. Paddison, L. 2022. COP15: More than 190 countries sign landmark agreement to halt the biodiversity crisis. CNN.
4. Isbell, F. et al. Expert perspectives on global biodiversity loss and its drivers and impacts on people. *Frontiers in Ecology and the Environment*, 2022; 21(2), 94–103. https://doi.org/10.1002/fee.2536
5. Balmori, A. Electromagnetic radiation as an emerging driver factor for the decline of insects. *Science of the Total Environment*, 2021; 767: 144913. For additional details see www.emfsafetynetwork.org and https://www.americansforresponsibletech.org
6. Ivey, M.R. et al. Reduced vertebrate diversity independent of spatial scale following feral swine invasions. *Ecology and Evolution*, 2019; 9: 7761–7767. https://doi.org/10.1002/ece3.5360
7. Feral Dogs – Wildlife Damage Management. Article available at: https://wildlife-damage-management.extension.org
8. Bergman, D. et al. *Dogs gone wild: Feral dog damage in the United States*. USDA National Wildlife Research Center – Staff Publications, 2009, 862. Article available at: https://digitalcommons.unl.edu/icwdm_usdanwrc/862
9. Belsare, A. and Vanak, A.T. Modelling the challenges of managing free-ranging dog populations. *Scientific Reports*, 2020; 10: 18874. https://doi.org/10.1038/s41598-020-75828-6
10. Doherty, T.S. et al. The global impacts of domestic dogs on threatened vertebrates. *Biological Conservation*, 2017; 210(Part A): 56–59. https://doi.org/10.1016/j.biocon.2017.04.007
11. Trouwborst, A. et al. Domestic cats and their impacts on biodiversity: A blind spot in the application of nature conservation law. *People Nature*, 2020; 2: 235–250. https://doi.org/10.1002/pan3.10073
12. Lepczyk, C.A. et al. A global synthesis and assessment of free-ranging domestic cat diet. *Nature Communications*, 2023; 14: 7809. https://doi.org/10.1038/s41467-023-42766-6
13. Hira, M. et al. *Toxoplasma gondii* IgG serointensity is positively associated with frailty. *Journals of Gerontology: Series A*, 2023; 79(3): glad 228. https://doi.org/10.1093/gerona/glad228
14. Flegr, J. and Horáček, J. Negative effects of latent toxoplasmosis on mental health. *Frontiers in Psychiatry*, 2020 Feb 18; 10: 1012.
15. Lundgren, E.J. et al. Functional traits – not nativeness – shape the effects of large mammalian herbivores on plant communities. *Science*, 2024; 383: 531–537.
16. Anderson, A.K. et al. Partial COVID-19 closure of a national park reveals negative influence of low-impact recreation on wildlife spatiotemporal ecology. *Scientific Reports*, 2023; 13: 687.

17. *Tennessee forests and woodlands.* Article available at: https://www.nature.org/en-us/about-us/where-we-work/united-states/tennessee/stories-in-tennessee/saving-a-forest-saving-the-earth/ (Accessed Oct 17, 2024).
18. Sánchez-Bayo, F. and Wyckhuys, A.G.K. Worldwide decline of the entomofauna: A review of its drivers. *Biological Conservation*, 2019; 232: 8–27.
19. Chief Seattle. *Letter to all.* California State University – Northridge. Article available at: https://www.csun.edu/~vcpsy00h/seattle.htm
20. Stone, C.D. *Should Trees Have Standing? Law, Morality, and the Environment.* Third edition. Oxford, Oxford University Press, 2010.
21. Himrod, C. Land Trust Alliance applauds introduction of Grasslands Act. July 27, 2022. Land Trust Alliance. Article available at: https://landtrustalliance.org/newsroom/press-releases/land-trust-alliance-applauds-introduction-of-grasslands-act
22. The Mississippi River Watershed Project: Article available at https://www.mvr.usace.army.mil/About/Offices/Programs-and-Project-Management/District-Projects/Projects/Article/1173148/upper-mississippi-river-
23. Barth, B.A. Modern Farmer Conversation: The Wisdom of Wes Jackson. March 3, 2017. *Modern Farmer.* Article available at: https://modernfarmer.com/2017/03/wes-jackson-the-land-institute-kernza/
24. North American Wetlands Conservation Act. Ducks.org. Article available at: https://www.ducks.org/conservation/public-policy/nawca-north-american-wetlands-conservation-act
25. Citation from SOWB2022_EN_compressed.pdf (birdlife.org) https://www.birdlife.org/papers-reports/state-of-the-worlds-birds-2022/
26. Cornell Lab. *Bring birds back.* Article available at: https://www.birds.cornell.edu/home/bring-birds-back/
27. Uhlfelder, E. *Bloody skies: The fight to reduce deadly bird-plane collisions.* November 8, 2013. National Geographic. Article available at: https://www.nationalgeographic.com/adventure/article/131108-aircraft-bird-strikes-faa-radar-science
28. Urbina, I. The crimes behind the seafood you eat. October 9, 2023. *National Geographic.* https://www.newyorker.com/magazine/2023/10/16/the-crimes-behind-the-seafood-you-eat
29. Meiling, B. *Would you eat lab-created fish? This startup is carving new path in 'alt-meat' industry.* Phys.org. May 22, 2019. Article available at: https://phys.org/news/2019-05-lab-created-fish-startup-path-alt-meat.html
30. Compassion in World Farming. *Rethink fish.* Article available at: https://www.ciwf.org.uk/our-campaigns/rethink-fish/#
31. Petersen, L.R. Epidemiology of West Nile virus in the United States: Implications for arbovirology and public health. *Journal of Medical Entomology*, 2019; 56: 1456–1462. https://doi.org/10.1093/jme/tjz085
32. Kolby, J. To prevent the next pandemic, it's the legal wildlife trade we should worry about. May 7, 2020. *National Geographic.* Article available at: https://www.nationalgeographic.com/animals/article/to-prevent-next-pandemic-focus-on-legal-wildlife-trade
33. Messenger, A.M. et al. Reverse zoonotic disease transmission (zooanthroponosis): A systematic review of seldom-documented human biological threats to animals. *PLOS ONE*, 2014 Feb 28; 9(2): e89055. https://doi.org/10.1371/journal.pone.0089055
34. Dunay, E. et al. Pathogen transmission from humans to great apes is a growing threat to primate conservation. *Ecohealth*, 2018 Mar; 15(1): 148–162. https://doi.org/10.1007/s10393-017-1306-1
35. The Columbian (Vancouver, WA) (tiered subscription model), December 23, 2023.
36. Hughes, B.B. et al. Top-predator recovery abates geomorphic decline of a coastal ecosystem. *Nature*, 2024; 626: 111–118.

4 A Beef about Beef
Reforms Long Overdue?

The continuing assaults documented in this review by the cattle industry on wildlands, biodiversity, indigenous peoples, and wildlife; contributions to climate change; and risks to consumers and public health call for full accountability and change.

Cattle ranchers in the United States and other countries have waged war on wolves and other predators for centuries, which ultimately disrupts the valuable ecological services these predators play in keeping herbivores such as deer and elk herds as well as forests, healthy – and indirectly benefiting public health (1).

This disturbing August 18, 2023, article by Christopher Ingraham entitled 'USDA Kills Hundreds of Minnesota Wolves to Protect Ranchers' underscores the perpetuation of this insanity (2). He writes:

> In 2022, there were 174 documented wolf deaths in Minnesota, according to the latest state Department of Natural Resources data. Of those, 142 were killed by a relatively obscure arm of the U.S. Department of Agriculture called the Wildlife Services Division. USDA officers kill more wolves in Minnesota than in all other states combined, according to the program's annual reports. That work is effectively a government handout to ranchers, who receive publicly funded protection for their privately held livestock. The ranchers also receive cash compensation from state taxpayers for their lost cattle, which in 2022 totaled $100,000 for 78 wolf predation claims, or an average of about $1,300 per claim. While individual ranchers can experience significant losses if wolves repeatedly target their cows, the overall impact on the state's cattle population is negligible. There are about 2.2 million cows in the state, according to USDA data. The five or six dozen documented and verified wolf kills each year amount to a few thousandths of 1 percent of the total population. But the USDA's actions in response inflict a steep toll upon Minnesota's wolves. The 142 kills amount to fully 5% of the state's estimated wolf population.

Objective research reviews of such lethal 'wolf control' conclude that killing wolves does more harm than good and can increase livestock losses. A 2014 report in the journal *PLOS One* found that livestock mortality increased as wolves were killed, possibly because the killings spurred surviving wolves to mate more often. 'The odds of livestock depredations increased 4% for sheep and 5–6% for cattle with increased wolf control – up until wolf mortality exceeded the mean intrinsic

DOI: 10.1201/9781003569787-4

growth rate of wolves at 25%. Possible reasons for the increased livestock depredations at <25% mortality may be compensatory increased breeding pairs and numbers of wolves following increased mortality. After mortality exceeded 25%, the total number of breeding pairs, wolves, and livestock depredations declined. However, mortality rates exceeding 25% are unsustainable over the long term. Lethal control of individual depredating wolves may sometimes be necessary to stop depredations in the near-term, but we recommend that non-lethal alternatives also be considered' (3).

Many Americans are still enchanted by the mythology of the pioneer's manifest destiny to conquer the Western frontier and the cowboys who, among others, committed genocide and ecocide, exterminating species – wolves and cougars – and peoples – the Anishinaabe – and almost annihilating the buffalo to fatten the coffers of the cattle barons.

The now-critical depletion of aquifers, degradation of public 'rangelands' by overstocked cattle, and destruction of prairie grasslands and draining of wetlands to raise feed for the livestock industry collectively amount to ecocide. This is compounded by chemical fertilizers that destroy soil microorganisms along with pesticides used on genetically engineered feed crops that have decimated beneficial insect and insectivore populations, from monarch butterflies to bats and barn swallows to swifts, and cause cancer in consumers

FIGURE 4.1 A postcard celebrating livestock supplanting buffalo, an illusion of progress.

and especially farm workers. Livestock fecal waste, loaded with potentially harmful, antibiotic-resistant bacteria, contaminates the air, and pollutes ground/surface and irrigation and drinking water. The highly publicized climate change contribution of methane from livestock to climate change and fecal pollution is the tip of the iceberg – or animal waste pile – created by the global livestock industry compounded by its hide-tanning and recycling subsidiaries.

The contribution of the U.S. beef industry to the climate crisis has been well documented, as well as the role of high beef consumption to various health problems, including cancer. Yet this industry continues to be heavily subsidized with public funds by government, which many analysts are now calling to be immediately terminated (4). According to one analysis, 'About 85% of public lands in the western USA are grazed by domestic livestock, and they influence climate change in three profound ways: they are significant sources of greenhouse gases through enteric fermentation and manure deposition; they defoliate native plants, trample vegetation and soils, and accelerate the spread of exotic species resulting in a shift in landscape function from carbon sinks to sources of greenhouse gases; and they exacerbate the effects of climate change on ecosystems by creating warmer and drier conditions' (5).

FIGURE 4.2 Beef cattle in a crowded Colorado feedlot with no shade or shelter.

To the cost of beef we should add the expenses of monitoring, containing, preventing, and treating the many diseases cattle can suffer from, some of which can infect us (6). These diseases can infect wildlife that can in turn infect cattle as with brucellosis and tuberculosis. As the human population increases along with livestock and demand for meat, there will be predicted increases in diseases harbored by wildlife infecting cattle and other farmed animals and more zoonotic diseases challenging the human populace (7).

Cattle are routinely treated with pesticides and other antiparasite drugs that are now a significant ecological and environmental concern. Since the 1950s, the U.S. Food and Drug Administration (FDA) has approved several steroid hormone drugs for use in beef cattle and sheep, including natural estrogen, progesterone, testosterone, and their synthetic versions, to boost productivity, along with various antibiotics in their feed, consumer risks notwithstanding. β-adrenergic agonists like Zilmax (now banned in the United States over animal well-being issues) and Optaflexx (ractopamine) are given to cattle and other farmed animals to make them leaner. But they are chronically stressed by this kind of neuro-psychotropic drug, making them hyperalert. China is genetically engineering massively double-muscled cattle and other animals for human consumption. The anti-inflammatory drug diclofenac, widely used until a recent ban, was highly toxic to India's vultures who were consuming contaminated cattle remains and were almost exterminated across the subcontinent.

According to the Humane Farming Association (HFA),

> The Livestock Indemnity Program (LIP) is like an insurance policy for livestock producers where the taxpayer pays the premium rather than the livestock producer. When animals die in harsh weather, the producer files a claim with the USDA. Under LIP, adverse weather events include, but are not limited to, earthquakes, hail, lightning, tornadoes, tropical storms, winter storms, hurricanes, floods, wildfires, extreme heat, and extreme cold. Compensating producers for dead livestock without ever requiring adequate shelter, shade, or other safeguards is a disincentive to farmers and ranchers to take the necessary steps to provide their animals with protections from inclement weather. Figures obtained under the Freedom of Information Act indicate that between 2013 and 2017, LIP issued payments of $198,877,658 to farmers and ranchers for animal deaths due primarily to weather-related issues. Since that time, says USASpending.gov, another $157,283,183 was paid to producers under LIP. LIP benefits represent misuse of taxpayer dollars and government waste at its most flagrant. (8)

To top this off, a vast tonnage of discarded animal parts condemned for human consumption – designated as '4-D' meat – from dead, dying, debilitated, and diseased animals at the processing plants is profitably recycled into pig and poultry feed and pet foods.

Our historical consumption of meat does not justify the present quantity and frequency. Raising an animal or tracking and killing one to eat is generations away for most people around the world. They have others kill and dismember them in slaughtering facilities, where there is much animal terror and too often mishandling and, especially in the United States, unsafe conditions for workers, many of whom are immigrants (9).

According to one analysis, (10) the U.S. federal government spends $38 billion every year subsidizing the meat and dairy industries. Research from 2015 shows this subsidization reduces the price of Big Macs from $13 to $5 and the price of a pound of hamburger meat from $30 to the $5 we see today. The U.S. government has blocked attempts to have 'Country of Origin' labels on all beef products in the marketplace (11).

From 2008 to 2017 the Brazilian government invested $22.2 billion in the beef industry. In two extraordinary years – 2015 and 2016 – tax revenue from the meat sector fell short of the government benefits it received. While problematic on their own, such economic incentives encourage cattle producers to burn additional rainforest, mostly illegally, to make additional room for cattle ranching, destroying endangered species habitat in the process (12).

Soybeans are imported to the United States from Brazil to feed livestock and poultry. The United States imported a record-breaking amount of fresh beef from Brazil in 2022. In January 2022 alone, imports reached nearly 100 million pounds – a more than 500% increase relative to the same month a year earlier – with fresh beef accounting for 83 million pounds, according to the U.S. Department of Agriculture (USDA) (13).

Brazil was the largest beef exporter in the world in 2020 followed by Australia, the United States, India, and Argentina. Seven countries exported more than 1 billion pounds of beef in 2020: Brazil, Australia, the United States, India, Argentina, New Zealand, and Canada (14).

Family-owned, Minnesota-based multinational Cargill Co. is deeply involved in these extractive activities that call for ending deforestation and violation of indigenous people's rights in the Amazon. The Amazon has started to release carbon. In 2021, air samples revealed that the rainforest's carbon uptake has weakened, and some parts have become a source of carbon dioxide. 'What we were predicting to happen perhaps in two or three decades is already taking place,' says climate scientist Carlos Nobre. Large-scale deforestation is the most visible threat. But even intact forest is no longer as healthy as it once was because of climate change and the impacts of agriculture that spill beyond farm borders (15).

FIGURE 4.3 Extensive pre-desertification soil erosion from overgrazing cattle in Tanzania.

India, with a population of 1,425,775,850 with an estimated 224.3 million undernourished in 2019–2021, according to a UN report, is yet one of the world's largest exporters of beef (from water buffaloes). About 1.5 million metric tons of carcass weight equivalent (CWE) of beef and veal were exported from India in 2022. India is one of the remaining top four beef exporters, with steady exports expected in 2023. India exports large quantities of water buffalo meat (carabeef) to low-end markets in Indonesia and Malaysia. Alan Colchester, a professor of neurology at the University of Kent, and Nancy Colchester, writing in the medical journal the *Lancet*, proposed a theory that the most likely initial origin of bovine spongiform encephalopathy (BSE) in the United Kingdom (mad cow disease decimated the UK's cattle industry and also infected at least 200 consumers) was the importation from the Indian subcontinent of bone meal that contained Creutzfeldt–Jakob disease (CJD)–infected human remains (16).

It was certainly a crime against nature when the World Bank and other development agencies funded the construction of thousands of miles of fences to support Botswana beef barons, resulting in the deaths of millions of wildlife unable to migrate seasonally to find food and water (17). Domestic cattle that roam and breed freely and become a feral invasive species is an issue in several countries, especially in Australia and the United States (18).

Many of India's 5 million stray cattle are discarded males (castrated bullocks used as beasts of burden) or milk cows that are no longer productive. State governments run by the party of Hindu Nationalist Prime Minister Narendra Modi have tightened laws to protect cows, making it harder for farmers to sell for slaughter cattle that are no longer giving milk (19).

There are safer and more sustainable alternative sources of nutrients, including microbial proteins, that need to be utilized for the common good and the good of the commons (20). The social, public health, and environmental costs of the entire global beef industry and associated inhumane concentrated animal feeding operations (CAFOs) for dairy cows, pigs, and poultry should be ethically unacceptable in any civilized, democratic society.

U.S. banks' financing of meat, dairy, and feed corporations and the sizable climate impact of that financing is a significant issue. Between 2016 and 2023, 58 U.S. banks provided $134 billion in lending and underwriting to meat, dairy, animal feed, food processing, and agri-commodity corporations, according to a review by the Netherlands-based research group Profundo and U.S.-based environmental organization Friends of the Earth. More than half of the financing examined in the report came from just three major banks: Bank of America (NYSE: BSE), Citigroup (NYSE: C), and JPMorgan Chase (NYSE: JPM). The study, titled *Bull in the Climate Shop: Industrial Livestock Financing Sabotages Major U.S. Banks' Climate Commitments*, finds that U.S. banks' lending to meat, dairy, and feed corporations results in approximately 11% of the greenhouse gas emissions linked to the banks' financing. However, these corporations represent just a tiny fraction (0.25%) of the banks' portfolios – a 44× difference. In other words, this financing presents an outsized impediment to banks meeting their climate commitments (21).

A significant next step in helping secure a viable planet for generations to come is transitioning to organically certified, grass-fed, free-range cattle production when the livestock industry is supported by consumers and governments, utilizing the positive ecological services of these animals as well as sheep and goats, when in low and healthy numbers, in the restoration and maintenance of grasslands, steppes, and prairies around the world. A modeling study of introducing bison into natural habitats showed that the 'animal presence fundamentally changes the relationships between plants, microbes, and the environment. In turn, this leads to large changes in the amount of carbon captured and stored in ecosystems relative to conditions that exclude animals. Hence, animals can be allies in fighting climate change as a key addition to the growing portfolio of nature-based climate change solutions' (22). When herbivores such as deer browse vegetation, they stimulate plant growth and that creates a bigger carbon sink (23). Deer emit much less methane than sheep and far less than cattle on the same diets (24).

The vegan dream of no cattle, sheep, or goats being farmed would simply contribute to increased atmospheric carbon dioxide without the stimulation of plant

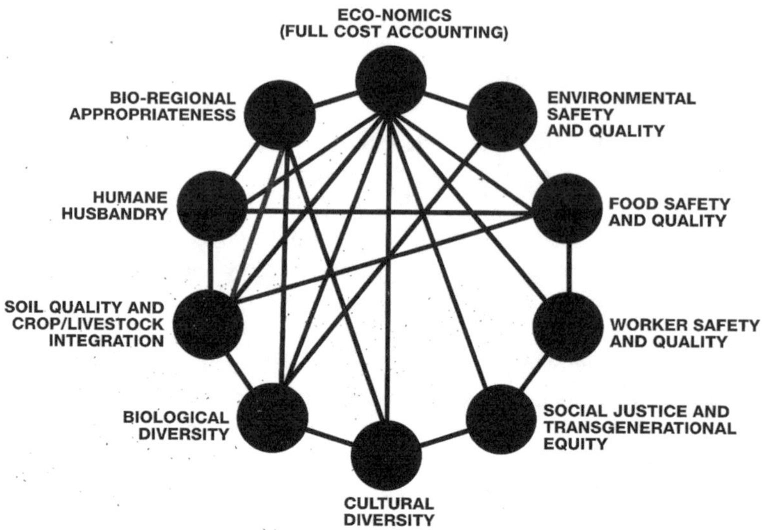

FIGURE 4.4 To farm without harm: A schematic showing the bioethical criteria and principles of organic, sustainable agriculture.

growth through animals grazing and browsing. Grasslands need to be restored and indigenous herbivores such as the buffalo re-introduced and, eventually, wolves, with the buffalo wolf being extinct. Consumption of vegetation also helps reduce the incidence of wildfires that are a significant factor in climate change. It would not be 'greenwashing' for ranchers to transition to optimal numbers of grazers, including cattle, sheep, and goats, and mixed grazing coupled with rotational grazing at low stocking densities, with eventual reintroduction of indigenous wild herbivores. Horizontal transfer of diseases such as chronic wasting disease, hemorrhagic fever, and brucellosis between wild and domestic herbivores will call for heightened veterinary involvement and the prudent use of available vaccinations. Greater vigilance and steps to prevent diseases from farmed animals infecting people are evident as public health and veterinary expertise come together to better prevent these zoonoses (25).

I have met with Western cattle ranchers and East African and South Indian cattle herders and witnessed how their animals are handled and slaughtered and grazing lands seriously degraded. All of us must change if we are to ever restore the healing connections of One Health for our One Earth. After all, none of the plants we eat can infect us with diseases (and many prevent us from disease and aging), while the animals we consume most certainly do.

In the next chapter the urgent need to protect predators such as the wolf and ending the livestock industry's war against these animals whose ecological services ultimately benefit our own health will be detailed.

REFERENCES

1. See documentation in Chapter 5: Wolves and Human Well-Being: Ecological and Public Health Concerns.
2. Ingraham, C. USDA kills hundreds of Minnesota wolves to protect ranchers' profits. MinnesotaReformer.com. August 18, 2023. Article available at: https://minnesotareformer.com/2023/08/18/usda-kills-hundreds-of-minnesota-wolves-to-protect-ranchers-profits/
3. Wielgus, R.B. and Peebles, K.A. Effects of wolf mortality on livestock depredations. *PLOS ONE*, 2014; 9(12): e113505. https://doi.org/10.1371/journal.pone.0113505
4. Vallone, S. and Lambin, E.F. Public policies and vested interests preserve the animal farming status quo at the expense of animal product analogs. *One Earth*, 2023; 6(9): 1213–1226. https://doi.org/10.1016/j.oneear.2023.07.013
5. Kauffman, J.B. et al. Livestock use on public lands in the Western USA exacerbates climate change: Implications for climate change mitigation and adaptation. *Environmental Management*, 2022; 69: 1137–1152. https://doi.org/10.1007/s00267-022-01633-8
6. Institutional Animal Care and Use Committee. *Zoonoses Associated with Cattle*. Washington State University, 2021. Article available at: https://iacuc.wsu.edu/zoonoses-associated-with-cattle/
7. Miller, R.S. 7 Farnsworth, M.L. Diseases at the livestock-wildlife interface: Status, challenges, and opportunities in the United States. *Preventive Veterinary Medicine*, 2013 Jun 1; 110(2): 119–132. https://doi.org/10.1016/j.prevetmed.2012.11.021
8. National Partnership for Women and Families 2023. *The Healthy Families Act*. Article available at: HFA-LIP-fact-sheet.pdf
9. Eisnitz, G.A. *Slaughterhouse: The Shocking Story of Greed, Neglect, and Inhumane Treatment inside the U.S. Meat Industry*. Prometheus Books, 2006.
10. Gillette, D. and Barge, W. *The True Cost of a Hamburger*. American Institute for Economic Research, April 20, 2022. Article available at: https://www.aier.org/article/the-true-cost-of-a-hamburger/
11. Carlson, P. 'Quit making money on the backs of American producers': USDA proposes U.S. beef labeling updates. *Drovers Farm Journal*, March 9, 2023. Article available at: https://www.drovers.com/news/ag-policy/quit-making-money-backs-american-producers-usda-proposes-us-beef-labeling-updates
12. Hofmeister, N. Brazilian taxpayers subsidizing Amazon clearing cattle ranches, study shows. *Mongabay*, May 26, 2020. Article available at: https://news.mongabay.com/2020/05/brazilian-taxpayers-subsidizing-amazon-clearing-cattle-ranches-study-shows/
13. Staheim, C. US imports from Brazil Continue. *Tri-State Livestock News*, May 12, 2023. Article available at: https://www.tsln.com/news/us-imports-from-brazil-continue/#
14. Peel, D. Major beef exporters and importers 2023. *Drovers Farm Journal*, December 12, 2022.
15. Grossman, D. 'We are killing this ecosystem': The scientists tracking the Amazon's fading health. *Nature*, August 24, 2023. Article available at: https://www.nature.com/immersive/d41586-023-02599-1/index.html
16. Colchester, A.C. and Colchester, N.T. The origin of bovine spongiform encephalopathy: The human prion disease hypothesis. *Lancet*, 2005; 366(9488): 856–861. https://doi.org/10.1016/S0140-6736(05)67218-2.
17. Williamson, D. *Botswana: Environmental policies and practices under scrutiny*. The Lomba Archives. IUCN Library System. 1994. Article available at: https://portals.iucn.org/library/node/21046

18. Thompson, J. The feds crack down on feral cattle. *High Country News*, March 2, 2023. Article available at: https://www.hcn.org/articles/south-landline-the-feds-crack-down-on-feral-cattle
19. Pathak, S. They're destructive: There are 5 million of them – and they're sacred. *National Geographic*, March 31, 2023. Article available at: https://www.nationalgeographic.com/animals/article/india-five-million-stray-cattle-sacred-hindu
20. Humpenöder, F. et al. Projected environmental benefits of replacing beef with microbial protein. *Nature*, 2022; 605: 90–96. https://doi.org/10.1038/s41586-022-04629-w
21. *Bull in the climate shop: Industrial livestock financing sabotages major U.S. banks' climate commitments.* Article available at: https://foe.org/resources/bull-in-the-climate-shop/
22. Rizzuto, M., Leroux, S.J., and Schmitz, O.J. Rewiring the carbon cycle: A theoretical framework for animal-driven ecosystem carbon sequestration. *Journal of Geophysical Research: Biogeosciences*, 2024; 129: e2024JG008026. https://doi.org/10.1029/2024JG008026
23. Malhi, Y. et al. The role of large wild animals in climate change mitigation and adaptation. *Current Biology*, 2022; 32: 181–196. https://doi.org/10.1016/j.cub.2022.01.041
24. Swainson, N.M. et al. Comparative methane emissions from cattle, red deer and sheep. *Proceedings of the New Zealand Society of Animal Production*, 2008; 68: 59–62.
25. Stel, M., Eggers, J., and Alonso, W.J. Mitigating zoonotic risks in intensive farming: Solutions for a sustainable change. *EcoHealth*, 2022; 19: 324–328. https://doi.org/10.1007/s10393-022-01605-8

5 Wolves and Human Well-Being

Ecological and Public Health Connections

Predators play a vital role in maintaining the biodiversity and associated health of aquatic and terrestrial ecosystems. Some have been long persecuted. In this chapter the role of wolves in providing these beneficial 'services' is reviewed with reference to the gray wolf population east of the Rocky Mountains in the United States.

I have studied wolf behavior and development, raised and lived with captive wolves, and have done in-field research on wild canids. Therefore, I feel that I have some authority to speak for them to see their intrinsic rights recognized and the sanctity of their existence, intrinsic value, and contribution to healthy ecosystems respected and protected from human encroachment and recreational and commercial exploitation.

Some anthropologists and evolutionary biologists contend that wolves helped humans evolve, through close observation and mimicry of the wolf, into followers and cooperative hunters of game species. This gives credence to such aboriginal aphorisms as 'The wolf is my brother' and 'Wolves made us human.' The systematic extermination of the wolf began when inadequately protected settlers' livestock were attacked and killed by wolves, in part because their indigenous prey had been displaced by human encroachment and the numbers reduced by hunters.

Lupophobia, the fear of wolves, probably has its origins in Europe during regional famines and wars, when starving wolves and dogs scavenged human corpses and were themselves crazed and ravaged by rabies and other diseases. Lupophobia is certainly not shared by indigenous Native American Indians or by a growing majority of non-native American citizens who oppose wolf hunting and trapping. The wolf is a species symbolic of a nation divided by a bipolar society that has yet to find unity of vision and values, ethics, and spirit.

Wolves play an integral role in maintaining the health and biodiversity of wildlife and ecosystems and, indirectly, livestock and public health. Recognition of this role, long acknowledged by Native American Indians as 'The good medicine of

DOI: 10.1201/9781003569787-5

FIGURE 5.1 Pack greeting ritual of wolves assembling before hunting.

the wolf,' calls for greater respect, protection, and increased numbers of wolves in appropriate habitats across North America and, in recognition of their sentience and family bonds, abolition of commercial and recreational trapping, snaring, and shooting. Current federal and state government initiatives, backed by diverse vested interests, are continuing to reduce the nation's existing wolf population, which is contrary to the directives of sound science, reason, and the public interest in 're-wilding' public lands and designated wilderness areas throughout the wolf's original range. The opening of public lands for private use, from logging and mining to ranching and other financially driven incursions, calls for vigilant conservation science and praxis, especially considering the adverse consequences of climate change, by responsible state and federal authorities.

DEER 'MANAGEMENT' = FARMING THE WILD

State wildlife management practices directed to maximize deer numbers for recreational hunters and rural America's virtual extermination of the wolf over the past two centuries, coupled with forest management practices and agricultural expansion indirectly providing feed for deer, and the encroachment of real estate housing developments with deer-attracting gardens and vegetation in municipal parks, have had unforeseen consequences associated with high white-tailed deer numbers, as well as elk in western states. Two of these unforeseen consequences concern public health and potential harm to the livestock industry, *which a higher population of wolves across the United States would do much to rectify.*

According to the Minnesota Department of Natural Resources (DNR), 'After the young (fawns) are born each spring, there are between 900,000 and 1,000,000 [white-tailed] deer in Minnesota. The hunting season is important to keep the deer population from getting too large. Each year, Minnesota hunters harvest between 150,000 and 200,000 deer.' Concern was voiced in 2014 that Minnesota's deer population had been reduced excessively by too many DNR hunting permits being sold.

Hunters seek out the healthiest deer, some seeking trophy antler-bearing bucks, while most go after young deer for the venison. A seasonal hunt eliminating almost one-quarter of the deer population means starvation for wolf in deer-hunted zones at the start of winter. This probably increases their predation on livestock. Increasing deer hunting quotas to better regulate deer numbers is not a biologically appropriate response, even though it is a multibillion-dollar source of revenue for states and equipment suppliers.

DISEASE CONCERNS

Wolves prey on deer year-round, taking healthy deer as well as those slowed and weakened by injury and disease. Their predation helps reduce the negative impact deer cause by browsing and destroying young trees. They also play a significant role in controlling diseases carried by deer, notably the prion-causing *chronic wasting disease* (CWD). This disease, which is carried by infected deer showing no evident signs of illness initially (1), also affects mule deer, elk, and moose. It is now spreading across the United States and Canada. Gilch et al. (2) summarize their review:

> Chronic wasting disease (CWD) is a prion disease of free-ranging and farmed ungulates (deer, elk, and moose) in North America and South Korea. First described by the late E.S. Williams and colleagues in northern Colorado and southern Wyoming in the 1970s, CWD has increased tremendously both in numerical and geographical distribution, reaching prevalence rates as high as 50% in free-ranging and >90% in captive deer herds in certain areas of [the] USA and Canada. CWD is certainly the most contagious prion infection, with significant horizontal transmission of infectious prions by, e.g., urine, feces, and saliva. Dissemination and persistence of infectivity in the environment combined with the appearance in wild-living and migrating animals make CWD presently uncontrollable, and pose extreme challenges to wild-life disease management. Whereas CWD is extremely transmissible among cervids, its trans-species transmission seems to be restricted, although the possible involvement of rodent and carnivore species in environmental transmission has not been fully evaluated.

Wolves are, however, probably immune to CWD (3). But if these prions mutate and cross the species barrier and affect livestock, especially since prions have now been found in plants consumed by deer and in agricultural crops consumed by livestock and humans, the consequences could have devastating economic consequences for the livestock industry. This could mirror the 'mad cow' disease–causing prion debacle in the UK, which led to mass slaughter and export bans to protect consumers of cattle infected with this form of spongiform encephalopathy from acquiring the debilitating and eventually fatal brain degeneration called Creutzfeldt–Jakob disease. An estimated 1 in every 2,000 people in the UK tests positive for the protein linked to the human version of mad cow disease (4), and alarmingly, prions have been found to have the ability to develop resistance to therapeutic drugs (5).

Prions have been found in crops and vegetation consumed by deer (6,7). To help control this disease, the spreading of which is facilitated when deer congregate in a small area to graze and browse, state deer management authorities should prohibit all 'recreational feeding' where land owners plant forage and put feed out to attract deer.

Although cases remain uncommon, the incidence of Creutzfeldt–Jakob disease is up 85% in the United States and 117% in Wisconsin, according to health officials, while CWD is spreading among deer, and some speculation has emerged that the trends could be linked. Macaque monkeys that ate venison from deer with CWD subsequently contracted CWD, researchers in Canada reported (8).

In October 2013, it was reported that cattle in Wisconsin were diagnosed with contracting insect-borne deer *epizootic hemorrhagic disease* (9), a first, and a warning for the livestock industry of the need to better monitor deer numbers and disease risks, especially CWD. Minnesota's 460 deer and elk farms, regulated by the state board of animal health, raise animals for their meat, antler velvet, and fenced-in trophy hunts and are a not-infrequent source of CWD. They pose a challenge to the state DNR to prevent the spread of this disease into the wild deer population.

Deer, elk, and buffalo ranchers across the country may put wildlife and livestock at risk when unwittingly keeping animals infected with CWD and other diseases communicable to ungulates, wild and domestic, including *brucellosis* (Bang's disease), *Johne's disease*, and *bovine tuberculosis. Eastern equine encephalitis* (EEE) is a fatal viral disease of horses that can infect a variety of avian and mammalian species but seldom causes clinical disease. In rare cases, white-tailed deer can be infected, and the disease is fatal in affected animals.

Organic farmers and environmentally conscious ranchers have long recognized the role of predators and other wildlife species in helping to preserve healthy ecosystems around and within their lands. An ecological approach to wildlife management and predator control policies and practices, moving away from what amounts to wildlife farming for recreational sports hunters of deer, elk, and other selected species, is called for now. In the final analysis, the wolf, long reviled by cattle and sheep ranchers and seen as a competitor to be exterminated by many deer hunters, may be the ultimate savior of America's livestock industry and of healthier deer and other wildlife numbers if the prion-induced CWD and other deer diseases communicable to livestock and other species, including humans, is to be stemmed biologically through wolf predation. This means more wolves in deer and elk habitat (as well as cougars, bears, and lynx).

LYME DISEASE

The Centers for Disease Control and Prevention documented over 30,000 cases of tick-borne *Lyme disease* in humans in 2012 (other estimates are as high as 300,000), a disease harbored by rodents and deer and which wolves, coyotes, foxes, and other carnivores can play an indirect role in helping to control (10,11). Deficient populations of small and large predators will mean higher numbers of Lyme and other tick-borne disease-carrying rodents and deer, and therefore a greater probability of human infection. These concerns underscore the need for a revolution in state and federal wildlife and natural resource management. The adoption of principles and practices that enhance biodiversity and healthy ecosystems – domestic, commercial, and wild – with greater protection for wolves as an integral aspect of a more enlightened and science-based approach to a better environment for all is the core principle of One Health. This is now being embraced worldwide by medical, veterinary, and other authorities and agencies. The North American white-tailed deer population, estimated at only 300,000 a century ago and now standing at 30 million (12), is an indicator of eco-dystopia, calling for concerted state and federal attention from the holistic perspective of conservation medicine, which must include increased predation by indigenous carnivores.

Concerning the rising incidence of Lyme disease in humans, Sam R. Telford (13) asserts that reducing deer numbers in many regions is called for but is opposed sociopolitically. Even though deer do not transmit Lyme disease to ticks, encouraging large deer populations for the hunting industry will mean an increase in adult female ticks feeding on deer. With fewer deer, there would be fewer larval and nymphal-stage ticks being born to feed on small-to-medium animals, such as mice, chipmunks, shrews, or birds infected with Lyme disease, thus lowering human risk.

The science of restoration ecology/ecological medicine clearly needs local and national political and economic support if these wildlands are to continue to exist and healthful biodiversity secured by reintroduction of apex predators such as the wolf. These forest lands provide our drinking water and many other sustainable natural resources, including botanical products discovered to treat cancer and other diseases, as well as provide an essential carbon sink to help combat global warming and climate change. In their totality and vital presence, they also enrich, inspire, and inform our souls. Where there are wolves, there is hope. As the deer need the wolves, so the rivers need the trees. As wolves learn the ways of the deer for the good of pack, so humans must learn the ways of the wolf for the good of all. The documented enhancement of ecological health reflected in greater biodiversity following the reintroduction of wolves into Yellowstone National Park (14) is a stellar example of the key role of the wolf in optimizing ecosystems and natural wildlife management, which humans have been unable to accomplish in the absence of this essential predator.

LIVESTOCK PREDATION

Killing wolves to reduce their numbers is not an effective wildlife management tool, nor can it be justified to protect livestock from wolf predation, according to one detailed analysis. The authors assert that

> [p]redator control and sport hunting are often used to reduce predator populations and livestock depredations, but the efficacy of lethal control has rarely been tested. We assessed the effects of wolf mortality on reducing livestock depredations in Idaho, Montana, and Wyoming from 1987–2012, using a 25-year time series. The number of livestock depredated, livestock populations, wolf population estimates, number of breeding pairs, and wolves killed were calculated for the wolf-occupied area of each state for each year. We found that the number of livestock depredated was positively associated with the number of livestock and the number of breeding pairs. However, we also found that the number of livestock depredated the following year was positively, not negatively, associated with the number of wolves killed the previous year. The odds of livestock depredations increased 4% for sheep and 5%–6% for cattle with increased wolf control—up until wolf mortality exceeded the mean intrinsic growth rate of wolves at 25%. Possible reasons for the increased livestock depredations at ≤25% mortality may be compensatory increased breeding pairs and numbers of wolves following increased mortality. After mortality exceeded 25%, the total number of breeding pairs, wolves, and livestock depredations declined. However, mortality rates exceeding 25% are unsustainable over the long term. Lethal control of individual depredating wolves may sometimes be necessary to stop depredations in the near-term, but we recommend that non-lethal alternatives also be considered. (15)

WOLVES SUFFERING MANGE AND OTHER DISEASES

Wolf numbers and population recovery resilience can be especially undermined periodically by outbreaks of the contagious skin disease called mange or scabies. It can become endemic – literally embedded in a population. Factors such as poor nutrition and stress (notably, a quarter million deer hunters taking one-third of their main food source every year, the white-tailed deer, now exacerbated by 6,000 wolf hunters and trappers); also, concurrent disease (such as hookworms) make animals susceptible to this contagious parasitic skin disease. It can spread rapidly within packs sharing the same resting-up and denning areas. Reported public sightings and photographing of emaciated wolves with little fur on their bodies and extensive bald patches near Duluth, relayed to the DNR, indicate that the wolves of this part of the state are sick and suffering. For them to be easily seen from the roadside indicates that these normally shy and elusive animals are in dire straits from this disease, not entirely of their own making or some 'natural' occurrence. Sick animals indicate sick ecosystems.

The parasite responsible for this disease, which can affect many species, including humans, *Sarcoptes scabiei*, causes intense and distressing itching, often leading to secondary bacterial and fungal skin infections and extensive alopecia due to hair follicle damage. Infested wolves with little or no insulating fur left on their bodies have been known to seek warmth and shelter in outbuildings in rural areas and will die of cold exposure when severely infested and lacking adequate coats. During the earlier era of America's 'war to exterminate predators,' Montana's state legislature passed a law in 1905 *requiring veterinarians to introduce sarcoptic mange into the wild canine population as part of their extermination efforts*. This disease persists to this day throughout the region, according to Dan Flores, in his book *Coyote America* (16), whose dramatic increase in numbers over the past decade or so was believed to be responsible for the equally dramatic decline in red foxes and for the spread of mange to wolves.

An evaluation of diseases afflicting Minnesota's wolves (17), Minnesota DNR's Michelle Carstensen et al. (see PDF Wildlife Health Program – Minnesota Department of Natural Resources files.dnr.state.mn.us/publications/wildlife/research2010/health.pdf) makes no mention of mange, only the percentage of wolves testing positive for various diseases identified in blood serum samples. Some of these findings raise serious concerns as to the long-term viability of the gray wolf in the western Great Lakes region, when the probable mortality and morbidity rates associated with disease are added to the numbers of wolves killed legally and illegally by trappers and hunters. In this study of the serum of 206 wolves, 75.4% tested positive for canine parvovirus, 76.1% for canine adenovirus, 16.4% for canine distemper virus, 2.8% for equine eastern encephalitis, 13.7% for West Nile virus, 9.6% for heartworm, 94.5% for Lyme disease, and 83.6% for neosporosis. Nine of 62 fecal samples tested positive for hookworm.

Echinococcosis is a parasitic tapeworm infestation in wolves and other carnivora, the eggs of which are passed out in the feces and, when ingested by deer, moose, and other cervids as well as livestock and humans, form cysts in various body organs. This is one significant public health issue that calls for greater vigilance by wildlife managers, handlers, and hunters and raises the question of interventive treatment of wolf packs with antiparasitic bait (18). Without adequate monitoring to estimate the severity of mange, Lyme disease, neosporosis, and other periodically devastating wolf diseases such as parvovirus and canine distemper which can decimate wolf litters – and which are red flags indicating the population is under stress/duress from multiple factors that need to be addressed by wildlife managers – how can any state determine accurate annual kill quotas for recreational and commercial killing?

One does not need to be a wildlife scientist or 'expert' to surmise that the impact of humans has enormous effect on the health and well-being of any wild species – especially on a complex, intricate, 'social' pack species like the wolf. When their environment is dramatically changed by thousands of insurgent hunters and trappers, the additional stress of being targets themselves is likely to exponentially exacerbate the stress–disease connection. No one can accurately quantify this, and 'scientific evidence,' if any, comes after the fact, too late to prevent their suffering and demise.

Efforts to protect indigenous wolf clans or cultures need to be redoubled, along with their restorative reintroduction into those parts of their original range where there is a viable population of deer and other wild prey for them. Cattle ranchers and other stakeholders can establish cooperative programs consonant with wolf protection and reintroduction/'re-wilding' – from adopting non-lethal methods of predator control to re-afforestation. The primary forest habitat of both the wolf and cougar, or American lion, is in dire straits across much of North America, notably due to:

- Climate change
- Invasive species (harmful insects, tree diseases, noxious and inedible plants)
- Habitat degradation by livestock overgrazing
- Overbrowsing by state-managed deer and other 'game' species
- Relentless human encroachment from commercial recreational and real estate developments such as mining, fracking, damming, and logging
- Acid rain contaminated with industrial and incinerator poisons like lead, mercury, cadmium, and dioxins and with pesticides used extensively by the agricultural and timber industries

Kiowa Native American and Pulitzer Prize winner, the late N. Scott Momaday, in his book *Earth Keeper*, makes this very clear, stating, 'We humans must revere the earth, for it is our wellbeing. If we treat the earth with kindness, it will treat us kindly' (19). Biologically, in our evolution, such love, including philopatry, our native attachment to place, familial bonds, and biophilia, were essential to our survival.

In realizing our biological and ecological affinities and relationships with other species – plant, animal, and microorganism (many of which nourish us and without which we would not survive) – we discover the existential roots of our spiritual and ethical connections of interdependence, symbiosis. The continued slaughtering of wolves and other fellow creatures, wild and domesticated, is a *moral injury* to all who care and must end; otherwise, we will bear witness to the extinction of ever more species and of our own humanity.

The revenue-generating belief that 'wildlife must pay its own way' adds up to a kind of death tax from the sale of hunting licenses. Taking the wolf off the Federal Endangered Species Protection List and putting wolves under state management is a money-driven political ploy creating more jobs and income, as well as pandering to ranchers and hunters' interests. Promoting the wolf as a valuable trophy species is another money-making policy, even with a lottery system of winning permits to kill. Revenues generated for hunt outfitters and equipment manufacturers serve as an additional economic and political incentive for state authorities to promote the killing of wolves and other wildlife as a recreational sport. This is in total disregard of a more ethical, scientific, and ecological approach to wildlife and habitat stewardship that does not give primacy to consumptive, exploitive, and predatory human interests and values.

Now there is scientific documentation that provides more proof on what we have long known: Legal wolf killing through public hunts leads to uncontrollable illegal wolf killing (20). Compensating livestock owners who practice non-lethal methods of predator control, as with guard dogs, for veterinary-verified wolf kills would be preferable to wolf extermination and a step toward coexistence analogous to the human–bear conflict. While providing feed for bears is a humane initiative with ever more forest fires, droughts, and lack of wild foods (21), the supplemental feeding of other species, such as elk and deer, could put these animals at risk. Putting out hay for elk when food is scarce in the winter months leads to the formation of large congregations, which can facilitate the spread of CWD. This calls for alternative spreading and distribution of such supplemental foods, since CWD could decimate elk herds (22).

I interviewed a cattle herder while studying predators in the Indian jungle in the 1970s (23) who told me he accepted some losses as 'paying his rent' to the tiger and other predators whose habitat he was exploiting. Japanese grain farmers once worshiped wolves at shrines and left food offerings near their dens, beseeching the elusive canid to protect their crops from wild boars and deer. Talismans and charms adorned with images of wolves protected against fire, disease, and other calamities and brought fertility to agrarian communities and to couples hoping to have children (24). Farley Mowat, in his seminal book, *Never Cry Wolf*, concluded, 'We have doomed the wolf not for what it is, but for what we deliberately and mistakenly perceive it to be – the mythologized epitome of a savage ruthless killer – which is, in reality, no more than a reflected image of ourself' (25).

SCIENTISTS WARRANT MASS WOLF SLAUGHTER FOR CONSERVATION PURPOSES

While acknowledging the impacts of habitat loss due to human encroachment, recreation-associated disturbance, and climate change on vegetation and caribou health, 34 scientists are endorsing the continued killing of wolves to 'save' the caribou because their numbers in western Canada are finally beginning to increase due to the slaying of thousands of wolves. Evidently, the science of wildlife management is in an ethical vacuum with a serious scotoma when it comes to envisioning and adopting a more holistic approach to habitat protection and restoration rather than the expedient 'remedy' of killing predators (26).

QUALITIES AND VIRTUES WOLVES AND HUMANS SHARE

While people in many states where there are still wolves are debating the pros and cons of hunting and trapping them, and some are seeking the delisting of the wolf from protection under the federal Endangered Species Act still in place in some states, understanding the nature of wolves may help us reach a societal consensus of unquestioned respect and continued protection for the following reasons. They are based on various in-field reports and my own research of socialized wolves, their behavior, development, and communication.

- They care for their young.
- They communicate cooperatively to survive.
- Offspring learn obedience and allegiance. They have devoted parents, older siblings, cousins, aunts, and uncles who nurture and educate them to be courageous but cautious, taking no risks.
- Adults bring toys and various objects for the cubs to play with.
- Cubs learn to read and decipher scents and interpret sounds, sights, tracks, and terrain, which they must know to survive.
- Family pack members share food and defend their own.
- They groom each other, removing burrs in their fur and giving healing licks to wounds.
- They express their intentions and do not conceal emotions.
- They show patience, curiosity, insight, and foresight.
- They care for their injured, providing food, and search for the lost.

- They are loyal, caring for old pack members too infirm to hunt, and mourn their passing.
- They sing in chorus, harmonizing, and play together.
- They reaffirm and celebrate their social and emotional bonds when they enjoin in extended family reunions or howl across their domain.
- Like us, they play and roll in fresh snow and chase snowflakes.
- They perfume themselves by rolling in fragrant organic materials they find on the trail.
- They kill to live, but not for sport.
- They have amicable relationships with other species, such as foxes and ravens, who clean up their kills.
- Wolf cubs play with ravens, which fly over them with sticks, teasing the cubs to jump up and grab the sticks.
- Like us, they have sporting games such as mock hunting and fighting; they follow rules, have a sense of fair play, and play safely to avoid injury.

We share these essential, sociobiologically evolved qualities for survival with wolves, but we kill them for sport and invade their territories, then justify their

FIGURE 5.2 Emotional intelligence evident in the eyes of a wolf cub and human infant.

eradication. Unlike us, wolves naturally control their numbers and provide environmental services that sustain biological diversity and ecological health. But like us, they will war with and kill members of rival groups when food resources and territory are limited. They know hunger and starvation, sustain physical injuries from hunting large prey, and suffer various infectious diseases – often transmitted by free-roaming and feral dogs.

Our biological kinship, which parallels in convergent evolution with the wolf, reverberates with the spiritual kinship our wolf-derived domesticated dogs bestow on us as devoted companions and which we have yet to fully reciprocate. We thus need to honor the wolf in the process as an ancestral teacher of survival in our gatherer–hunter past that indigenous peoples around the world acknowledge today in myth and legend. Recognition of these similar attributes of wolves and humans is notable in several indigenous traditions, such as the Shinto of Japan. A good wolf and a bad wolf are seen in human nature and acknowledged as being in every child and are embraced with understanding, so the good wolf is nurtured for the common good and the good of the commons. This is also seen in the Cherokee tradition in the story *Beyond the Conflict of Inner Forces*, which addresses how best to treat these two wolves in our own nature (27). From the perspective of philosopher Mary Midgley, we have an ancient fear of the wild, with 'going wild' being a threat to 'civilization.' This is compounded by zoophobia and especially lupophobia, when, as Prof. Midgely states, 'The projection of human vices seems to serve the purpose of making people in settled society feel justified in killing these creatures for their own convenience. They are symbolically killing their own wildness' (28). We do indeed project, moralistically and judgmentally, those aspects of human nature that we would rather deny that are evident in wolves and other animals that kill to survive. Such animals are then despised and seen as evil.

The foundation of democracy is based on the concept of mutual acceptance, respect, and equal rights for all protected under the law. The divergent interests of various segments of society can undermine and even destroy this foundation if there is not an overarching, unified sensibility of respect for all life. That includes wolves.

For a democratic government and society to be sustainable, the 'personhood' of non-human animals such as wolves and whales should be recognized and protected under international law. Such a trans-species democracy would benefit humankind by virtue of the ecological services that these and other species provide and, in the process, contribute, under the environmental banner of One Health, to our own well-being.

It is notable that the national environmental group Nonhuman Rights Project in 2024 called on Gov. Spencer Cox to veto a bill that Utah lawmakers had passed prohibiting the use of legal personhood in cases involving various entities, including any member of a taxonomic domain that is not a human being. In contrast, legal personhood has been signed into law in New Zealand for the Whanganui River and the Urewera. Now Māori and Pacific leaders are proposing the same for whales at the UN as a further step to protect the environment.

WILDLIFE RESEARCH: BIOETHICAL BOUNDARIES AND VETERINARY SUPERVISION

Putting radio collars on wolves and other wildlife can put animals at risk. The major problems are higher mortalities and morbidity associated with capture stress and the physical and behavioral problems caused by the neck collars (29). The ethics and animal welfare implications of such devices are considerable (30). Ticks can lodge under these collars and transmit various diseases and cause extreme discomfort and secondary skin infection with possible flesh-fly maggot infestation.

There are several documented and many word-of-mouth accounts of chemically immobilized and otherwise restrained endangered species like the Asian elephant and African wild dog being severely injured, killed, or dying soon after capture and/or release. In some instances, there was an association with the animals being injected with untested and unapproved modified live-virus vaccines. In other instances, the injured or killed animal was a pregnant or nursing mother.

Experienced veterinary supervision is called for, especially when research biologists are loose in the field using drugs and vaccines on their animal subjects and applying various methods of capture and restraint, which may cause serious injury, capture myopathy, and even death.

The future of the world's first intercontinental cheetah introduction program is under question after it emerged that three relocated animals, and three of their cubs, died in the space of 8 months. The cheetah has long been extinct in India. Last September, India launched with great fanfare the ambitious program to introduce the South African cheetah (*Acinonyx jubatus*) into Kuno National Park. Radio tracking collars may have caused infections that killed two wild cheetahs in India's Kuno National Park, according to wildlife experts and veterinarians. Conservationist Yadvendradev Jhala, who helped oversee the cheetah relocation project, says high humidity may have caused skin irritation beneath the collars, and the cheetahs opened wounds when they scratched (31).

But the non-ionizing radiation associated with collars and implants for radio tracking involves very high frequency (VHF), ultra-high frequency (UHF), and global positioning system (GPS) technologies, including via satellites where platform terminal transmitters (PTTs) are used, as well as geo-locating capabilities using satellites, radio-frequency identification (RFID) chips, and passive integrated responder (PIT) tags, among others. This can be harmful to a variety of species, as has been documented (32).

FIGURE 5.3 Less interventive wildlife research methods are called for when radio collars on animals such as wolves can cause injuries and infection.

The American Cancer Society's posting (33), noting that the International Agency for Research on Cancer has identified radio-frequency exposure as a possible carcinogen, emphasizes that the U.S. Environmental Protection Agency and the U.S. National Toxicology Program have not yet formally classified radio-frequency radiation in terms of its cancer-causing potential.

Wildlife continues to be harassed, stressed, and subjected to these in-field risks so that tissue and blood samples can be taken (though DNA evidence can be obtained from feces and rubbing/marking areas), radio collars and even cameras fitted, and microchips implanted. The generation of more scientific data from such field research may help advance careers and engender more funding and give some substance to wildlife management schemes. But when the animals in question are put at risk and there are no in-place regulations and effective law enforcement to protect and restore their existing habitats and to extend the same to help minimize the accelerating loss of genetic biodiversity, then these wildlife researchers should cease and desist.

Such activities alone have nothing to do with wildlife conservation and, at best, give the false impression that something is being done, with the foreign presence alone being a deterrent to poaching. Yet from a bioethical perspective, the risks to the animals far exceed the immediate and foreseeable benefits. So, I appeal to all appropriate institutions, governmental and non-governmental, for-profit and not-for profit, to encourage alternative, non-invasive wildlife research and to cease funding and permitting any form of wildlife capture except for urgent veterinary and conservation-translocation reasons.

In the next chapter we will examine the essential links of animal rights, bioethics, and conservation and the negative impacts of human population growth and current agricultural practices that need to be addressed for the common good and the good of the commons.

REFERENCES

1. Tamgüney, G. et al. Asymptomatic deer excrete infectious prions in faeces. *Nature*, 2009; 461: 529–532. https://doi.org/10.1038/nature08289
2. Gilch, S. et al. Chronic wasting disease. *Topics in Current Chemistry*, 2011; 305: 51–77. https://doi.org/10.1007/128_2011_159.
3. Zhang, J. The nature of the infectious agents: PrP models of resistant species to prion diseases (Dog, Rabbit and Horses). In J.M. Verdier (eds.), *Prions and Prion Diseases: New Developments*. NOVA Science Publishers, 2012, Chapter 2, pp. 41–48.
4. Gill, O.N. et al. Prevalent abnormal prion protein in human appendixes after bovine spongiform encephalopathy epizootic: Large-scale survey. *British Medical Journal*, 2013; 347: f5675.
5. Berry, D.D. et al. Drug resistance confounding prion therapeutics. *Proceedings of National Academy of Science*, October 15, 2013; 110(44): E4160–E4169. https://www.pnas.org/doi/10.1073/pnas.1317164110
6. Johnson, C. et al. Uptake of Prions into Plants. In session Current Science of Chronic Wasting Disease: What Have We Learned in the Last 5 Years? August 9, 2013. Article available at: https://chronic-wasting-disease.blogspot.com/2013/09/uptake-of-prions-into-plants.html
7. Pritzkow, S. et al. Grass plants bind, retain, uptake, and transport infectious prions. *Cell Reports*, 2015 May 26; 11(8): 1168–1175.
8. Race, B. et al. Chronic wasting disease agents in nonhuman primates. *Emerging Infectious Disease*, 2014 May; 20(5): 833–837. https://doi.org/10.3201/eid2005.130778
9. Schultz, R. Rare disease reported in Wisconsin cattle. *Wisconsin State Journal*. October 3, 2013.
10. Levi, T. et al. Deer, predators, and the emergence of Lyme disease. *Proceedings of the National Academy of Sciences of the United States of America*, 2012 Jul 3; 109(27): 10942–10947.
11. Hofmeester, T.R. et al. Cascading effects of predator activity on tick-borne disease risk. *Proceedings of the Royal Society B*, 2017; 284: 20170453. https://doi.org/10.1098/rspb.2017.0453
12. Cambronne, A. *Deerland: America's Hunt for Ecological Balance and the Essence of Wildness*. Guilford, CT, Lyons Press, 2013.
13. Telford, S.R. Deer reduction is a cornerstone of integrated deer tick management. *Journal of Integrated Pest Management*, 2017; 8: 25. https://doi.org/10.1093/jipm/pmx024
14. Ripple, W.J. et al. Trophic cascades from wolves to alders in Yellowstone. *Forest Ecology and Management*, 2015; 354: 254–260.

15. Wielgus, R.B. and Peebles, K.A. Effects of wolf mortality on livestock depredations. *PLOS ONE*, 2014; 9(12): e113505. https://doi.org/10.1371/journal.pone.0113505
16. Flores, D. *Coyote America: A Natural and Supernatural History.* New York, Basic Books, 2016.
17. Carstensen, M. et al. A serosurvey of diseases of free-ranging gray wolves (*Canis lupus*) in Minnesota, USA. *Journal of Wildlife Diseases*, 2017;53:(3), 459–471.
18. Cerda, J. et al. *Echinococcus* spp. tapeworms in North America. *Emerging Infectious Diseases*, 2018; 24(2): 230–235. https://doi.org/10.3201/eid2402.161126
19. Momaday, N.S. *Earth Keeper.* New York, Harper Collins, 2020.
20. Oliynyk, R.T. Human-caused wolf mortality persists for years after discontinuation of hunting. *Scientific Reports*, 2023; 13: 11084. https://doi.org/10.1038/s41598-023-38148-z
21. Rogers, L.L. Does diversionary feeding create nuisance bears and jeopardize public safety? *Human-Wildlife Interactions*, 2011; 5(2): 287–295.
22. Cook, J.D. et al. *Evaluating Management Alternatives for Wyoming Elk Feedgrounds in Consideration of Chronic Wasting Disease.* U.S. Department of the Interior, U.S. Geological Survey, November 2023. Open-File Report 2023–1015 Version 2.0.
23. Fox, M.W. *The Whistling Hunters: Field Studies of the Asiatic Wild Dog* (Cuon alpinus). Albany, NY, State University of New York Press, 1984.
24. Walker, B.L. *The Lost Wolves of Japan.* Seattle, WA, Washington University Press, 2008.
25. Mowat, F. *Never Cry Wolf.* Toronto, Canada, McClelland and Stewart, 1963.
26. Lamb, C.T. et al. Effectiveness of population-based recovery actions for threatened southern mountain caribou. *Ecological Applications*, 2024; 34 (4): e2965. https://doi.org/10.1002/eap.2965
27. Author Unknown. *Beyond the Conflict of Inner Forces.* Article available at: https://www.awakin.org/v2/read/view.php?tid=927
28. Midgley, M. The Problem of Living with Otherness. In *The Myths We Live By.* pp. 164–168. Abingdon, Oxfordshire, Routledge, 2004.
29. Mech, D.L. and Barber, S.M. A Critique of Wildlife Radio-Tracking and Its Use in National Parks. A Report to the US National Parks Service, 2002. http://npshistory.com/publications/wildlife/radio-tracking-2002.pdf
30. Lininger, H. and Lininger, B. 2017. Unlocking the 'Virtual Cage' of Wildlife Surveillance. https://scholarship.law.duke.edu/cgi/viewcontent.cgi?article=1339&context=delpf
31. Pandey, G. July 19, 2023. Kuno cheetah deaths: Could radio collars be killing the big cats in India? BBC.com/news/world-asia-india 66241816
32. Manville, A.M. et al. Health and environmental effects to wildlife from radio telemetry and tracking devices – State of the science and best management practices. *Frontiers in Veterinary Science.* 2024; 11. https://www.frontiersin.org/journals/veterinary-science/articles/10.3389/fvets.2024.1283709
33. American Cancer Society. Microwaves, Radio Waves and Other Types of Radiofrequency Radiation, May 31, 2016. https://www.cancer.org/cancer/cancer-causes/radiation-exposure.html

6 One Health
Incorporating Animal Rights and Conservation

Humankind has not woven the web of life. We are but one thread within it. Whatever we do to the web, we do to ourselves. All things are bound together. All things connect.

–Chief Seattle, Duwamish

FIGURE 6.1 Web-making spiders are a visual metaphor of the interconnected 'seamless' web of life.

DOI: 10.1201/9781003569787-6

Advancing social acceptance of animal rights and conservation has been protracted by vested interests for several decades. Their incorporation into the philosophy and praxis of One Health is integral to making progress in the realms of public health and economic security that have also been limited by lack of attention to the critical issues of diet, population, regenerative agriculture, and loss of natural biodiversity. Progress in One Health should not be conceptually limited or diverted by the profit-driven promotion of ever more vaccines, pharmaceuticals, and pesticides. A more holistic and integrated approach to pest and disease management linking animal, environmental, and human health would reduce overreliance on vaccines, antibiotics, and pesticides in particular that can cause collateral harm to beneficial species and, ultimately, to public health. However, the integrated approach would not negate the value of such agents when used in accord with the precautionary principle.

Many wild and domesticated animal species have served us in myriad ways for millennia in times of peace and war: For food, clothing, fuel, labor, protection, hunting, herding, tracking, guiding, transporting, rescue, communication, companionship, entertainment, exhibition, competition, and education. More recently, animals have been assistance providers and co-therapists. Additional ways animals have served humanity is to test military weapons and to advance human health through biomedical testing, research, and creating animal models of disease and genetically engineering them to serve as organ donors. Species providing these services include bees and other insects, frogs, pigeons, hawks, cormorants, dolphins, elephants, rats, rabbits, cats and dogs, cattle, sheep, pigs, poultry, equids, and camelids.

As the human population increased, so did the rate of animal use for various culturally accepted and often financially driven purposes. It was not until the early 19th century that various forms of exploitation, involving mistreatment and suffering, began to be questioned and animal protection and welfare organizations established. But we have yet to fully acknowledge the empathy, trust, loyalty, and devotion other animals have given us without question since our earliest relationships with them.

According to Prof. I.J.H. Duncan,

> Up until the 17th century, philosophers regarded animals as being quite distinct from human beings; human beings had rationality whereas animals had none. This meant that animals had only instrumental value and could be used in any way that human beings desired. During the Enlightenment, philosophers started to realize that the distinction was not clear-cut; animals had some rationality Bentham (1823) pointed out that rationality was not the important factor; animals could suffer, and that was what mattered; animals had intrinsic value. Also, during the 19th century, as part of Darwin's theory of evolution by natural selection, it was seen that states

> of suffering and states of pleasure could also be adaptive. Although the foundation was now in place, the emergence of modern animal welfare science was delayed through the first 70 years of the 20th century by Behaviorism, which eschewed any consideration of subjective experiences. (1)

Behaviorism amounts to the mechanomorphizing' of animals as instinct-driven automatons to counter any anthropomorphizing of animal consciousness. I agree with Duncan that this spell was broken by two books. I was fortunate to know and work closely with the authors of these ground breaking books, one addressing the cruelties of animal factory farms and the other raising a legitimate, science-based issue of animal sentience. The books are *Animal Machines* by Ruth Harrison (1964) and *The Question of Animal Awareness* by Donald Griffin (1976).

Subsequently, several philosophers and attorneys argued the case for animals to have rights, but made little progress without the backing of veterinarians, ethologists, and other biological scientists who could provide evidence of species' interests, behavioral, socioemotional, and environmental needs, as well as physical and nutritional requirements. These are core bioethical principles of animal husbandry and humane stewardship that help ensure animals' health and well-being under the banner of duty of care, elements of which are now incorporated in animal welfare and anti-cruelty laws in various countries.

The Cambridge Declaration on Consciousness, publicly proclaimed in Cambridge, UK, on July 7, 2012, was a landmark contribution from the scientific community. Significant advances in understanding animal consciousness and sentience have been made since then (2).

From a One Health perspective, environmental health/protection, as advocated by ecologists and conservationists, is linked with animal rights, animal health, and well-being that mirror our own degree of physical and mental health and socioeconomic well-being. The mental health benefits of contact with nature and hearing birdsong are now being documented (3,4). All who oppose environmental and animal protection and rights may someday be found guilty of crimes against humanity by eroding the bioethical basis of One Health.

Books such as Rachel Carson's 1962 *Silent Spring*, Paul Erlich's 1968 *The Population Bomb*, Frances Moore Lappe's 1971 *Diet for a Small Planet*, and the Club of Rome's 1972 report *Limits to Growth* helped galvanize public concern and sparked several non-profit organizations dedicated to addressing these connected concerns. But they did not turn the tide. The world's population is expected to increase by nearly 2 billion persons in the next 30 years, from the current 8 billion to 9.7 billion in 2050 and could peak at nearly 10.4 billion in the mid-2080s (5).

According to Compassion in World Farming, approximately 70 billion farm animals are reared for food in the world each year. Two out of every three farm animals in the world are reared on a factory farm. In the United States about 99% are reared on a factory farm (6).

Diseases from animals to humans is not only a foreign problem. Industrial agriculture, the exotic pet trade, fur farming, and live animal markets in the United States pose significant risks for zoonotic disease spillover events, and the United States lacks a comprehensive strategy for mitigating the danger, according to a report from experts at Harvard Law School and New York University. An estimated 25 million birds pass through some 130 live markets just in the Northeast every year, and there is evidence of outbreaks of highly pathogenic avian influenza at live bird markets and that swine flu has spilled over to people at live animal markets in the past (7). Following this same line of evidential concern, all state fairs and agricultural shows involving live animals should be closed if not under full veterinary, public health, and animal welfare surveillance and limited in size and frequency. For similar reasons, all traffic and trade in live, domesticated, and wild animals should be prohibited except for legitimate conservation and animal health purposes. The lesson from the live animal 'wet' markets in Wuhan, China, from where the COVID-19 global pandemic is thought to have originated, is warning enough.

FIGURE 6.2 Live animal 'wet' markets in China and other countries are inhumane and a significant public health hazard that should be closed down.

Natural population controls for most species come through predation, starvation, and disease. We humans have virtually eliminated predators, at documented ecological cost, enjoying considerable food security with pesticides and disease preventives and treatments. The consequences of such 'progress' for the human species account for us reaching a population density that with current rates of resource consumption (and pollution) exacerbate climate change and the evident ecocide in the escalating loss of biodiversity, the keystone of ecological health. The politicization of birth control, supported by religious factions denying women's rights to prevent and terminate pregnancies, is on the same chauvinistic continuum that denies animal rights and eco-justice.

Clearly, family planning to reduce population growth and reduce the production and consumption of farmed animals go hand in hand with saving biodiversity and the collective contribution to the extinction and climate crises. Aside from the documented health benefits of plant-based/vegan diets that help prevent cancer and other health issues; Oxford University scientists have calculated the environmental consequences of our dietary choices under current UK climatic conditions. A plant-based diet would result in 75% fewer greenhouse gas emissions (i.e., lower carbon footprint) than those who eat more than 3.5 ounces of meat daily, with less harm to land, water, and biodiversity. The study found that, compared to meat-heavy diets, vegan diets resulted in 75% less land use, 54% less water use, and 66% less biodiversity loss (8).

This is yet another confirmation of the ethical imperative of reducing meat production and consumption and de-emphasizing reliance on animals as dietary staples, in accordance with the theory and praxis of One Health. From a veterinary perspective, those involved in the farm animal sector can contribute to the long-overdue transition to humane and sustainable agriculture. Transitional agriculture calls for a dramatic reduction in the numbers of animals being produced and harvested from the land and sea for human consumption and in using good land to raise feed for livestock and poultry.

Reclamation, restoration, and re-wilding, especially of wetlands and grasslands taken over by industrial agriculture and sheep and cattle ranching, would be more feasible with consumer support of humane and ecologically sensible dietary choices, i.e. plant-based, ideally organically certified, and less or no meat and other animal-derived foods. This would go far to help rectify climate change and reduce zoonotic and food-borne diseases. In restoring bioregionally indigenous plant and animal species, including predators, the ecology, environment, and public health would improve. Selected, locally adapted breeds of farmed animals could be conserved when integrated with wildlife, with guard dogs serving as needed, to help farmers transition to more sustainable and regenerative land and water use. The veterinary profession can lead the way in monitoring and preventing zoonotic diseases transmitted by domesticated and wild animals and help prevent the horizontal spread of diseases between the wild and the domesticated.

These initiatives, along with climate-controlled hydroponic and other enclosed food production systems and technologies, including cellular, for local food production are enlightened responses to unpredictable climate change and uncertain crop yields. Local floods and droughts of increasing frequency, intensity, and duration have made farming more and more challenging. We face massive forest fires and the loss of oxygen-producing, carbon-sequestering trees. The oceans are warming and acidifying, and we are losing oxygen-producing, carbon-sequestering plankton. Arctic warming is disrupting the jet stream, leading to a possible collapse of the climate-stabilizing Atlantic Ocean current, with increased freshwater input from ice melt. Due to all these factors, along with global petrochemical and sewage pollution of aquatic ecosystems that also influence climatic stability, we and all life on Earth face a challenging future.

Commercial fishing that depletes global fish stocks and threatens the livelihoods of indigenous coastal fishing communities, along with ecologically damaging commercial shrimp and salmon farming, all need to end. Whaling must cease. Public funding via government subsidies and the World Bank and other bank 'development' loans given to all animal-based and associated industries should be limited to enterprises supporting humane, sustainable food production and developing transitional, regenerative agriculture. Human resources, priorities, intelligence, and political will are now being tested as we all face a rapidly changing climate that may take millennia to stabilize.

In the summer of 2023 approximately 40 million people in Kenya, Somalia, South Sudan, and Ethiopia faced a hunger crisis on an unimaginable scale. Climate-related drought was one of the major causes of the crisis – together with conflict and high food prices. Reducing food waste is one of the biggest opportunities for lowering greenhouse gas emissions in the world. The establishment of a United Environmental Nations is beginning to crystallize as responsible nation-states come together to protect biodiversity; ensure food, fuel, and water security; and address associated humanitarian crises and internecine strife. However, this group has yet to address the core issues of population growth and raising billions of animals for food.

The physician's Hippocratic injunction, 'First do no harm,' is the bioethical principle of *ahimsa*, non-harming, applicable to all our relationships with other living organisms (9). The path of least harm to ourselves, to the environment, and to health-promoting natural biodiversity in striving to prevent diseases, pests, and predators calls for a more judicious use of safe vaccines, antibiotics, pesticides, bioremediation (including soil and gut microbiome restoration), and biocontrol (as by introducing specific pathogens or predators to control crop pests and invasive species). Overuse and misuse of pharmaceutical and petrochemical products continue to cause more harm than good when the precautionary principle is ignored,

FIGURE 6.3 Planting cover crops to prevent soil erosion and crop rotation rather than monocultures are keys to sustainable agriculture.

harming beneficial animals, plants, and microorganisms and spawning the evolution of chemical-resistant 'superbugs' and 'superweeds.'

In conclusion, the conceptualization of One Health enables us to transcend anthropocentrism so that the current Anthropocene era can evolve into the bioethically based 'Empathocene' era to prevent continued ecocide, treating nature simply as a resource and animals as objects or commodities. I envision this Empathocene era as the stage in human evolution based on ecological sensibility; sound science; integrative, holistic medicine; and respect for all life. Environmental justice recognizes the rights of all living beings to equal consideration. Such eco-democracy is a step forward for civil society, where public health and economies are failing and democratic governance is endangered by continued erosion of social justice around the world.

The ethical imperative – as well as enlightened self-interest – in changing our diets, will be addressed more extensively in the next chapter.

REFERENCES

1. Duncan I.J.H. Animal welfare: A brief history. In Hild S. & Schweitzer L. (eds.), *Animal Welfare: From Science to Law* (pp. 13–19), 2019. La Fondation Droit Animal, Éthique et Science.

2. Birch J. et al. Dimensions of animal consciousness. *Trends in Cognitive Sciences*, 2020; 24: 789–801. https://doi.org/10.1016/j.tics.2020.07.007
3. Nutsford, D. et al. An ecological study investigating the association between access to urban green space and mental health. *Public Health*, 2013; 127: 1005–1011.
4. Stobbe E. et al. Birdsongs alleviate anxiety and paranoia in healthy participants. *Scientific Reports*, 2022; 12: 16414.
5. United Nations. *Global Issues: Population*. https://www.un.org/en/global-issues/population#
6. Animal Matters. *Facts – Farm Animals*. https://www.animalmatters.org/facts/farm/
7. Linder, A. et al. *Zoonotic Disease Threats in the US Uncovered in a Comprehensive New Report*. Animal Law and Policy Program, Harvard Law School, 2023. https://animal.law.harvard.edu/news-article/animal-markets-and-zoonotic-disease
8. Scarborough, P. et al. Vegans, vegetarians, fish-eaters, and meat-eaters in the UK show discrepant environmental impacts. *Nature Food*, 2023; 4: 565–574. https://doi.org/10.1038/s43016-023-00795-w
9. Fox M.W. *Bringing Life to Ethics: Global Bioethics for a Humane Society*. Albany, NY, State University of New York Press, 2001.

7 Changing Our Diets for Our Health and the Earth's Sake

We are slow to accept that the human population, coupled with its non-sustainable industrial and agricultural practices, dietary habits, and other environmentally destructive activities, has become a planetary infestation for which the immediate solutions are in the realm of self-control, especially of our breeding and feeding behaviors. If the forces of natural selection with attendant disease, starvation, and suffering, as well as internecine strife and wars, regional and international, over access to and control of natural resources are to be somewhat alleviated for the good of future generations, then this generation must begin to squarely address these interconnected issues and implement the necessary changes. Some of these are identified in this review to give grounds for hope of a viable future.

Climate change with extreme weather events that impact food production and food security is an existential crisis many communities are now facing. The related issue of rising temperatures and humidity that make insect-borne diseases spread and preventing pandemics will be discussed in more detail in the next chapter.

The United Nations Food and Agriculture Organization (FAO) has completed the first-ever global assessment of the state of the planet's land resources, finding that a quarter of all farmland is highly degraded and warning that the trend must be reversed if the world's growing population (expected to reach 9 billion by 2050) is to be fed. In my estimation, this is not possible, regardless of the promises of agricultural bioscience and other new technologies that are being promoted as panaceas by various governments, allied corporations, and international banks. The projected need, according to this report, amounts to 1 billion tons more wheat, rice, and other cereals and 200 million more tons of cattle and other livestock products above what is being currently produced annually (1).

However, as the report stresses, most available farmland is already being farmed and in ways that decrease its productivity through practices that lead to soil erosion and wasting of water. This means that to meet the world's future food needs, a major 'sustainable intensification' of agricultural productivity on existing farmland will be necessary, according to the FAO report.

DOI: 10.1201/9781003569787-7

This report emphasizes that climate change, coupled with poor farming practices, has contributed to a decrease in productivity of the world's farmland. Approximately 25% of the world's farmland is now 'highly degraded,' due to soil erosion, water degradation, and biodiversity loss. Another 8% is moderately degraded, while 36% is stable or slightly degraded and 10% is ranked as improving.

The opposition to organic, sustainable, and socially just agriculture by multinational corporations, touting their patented, genetically engineered seeds and pesticides under the false pretext of helping alleviate world hunger, is seen by many analysts as nothing short of criminal behavior. Such corporate hegemony and colonialism, aided and abetted by allied governments and aid and development institutions like the World Bank, have contributed significantly to the global food crisis that we face today, most especially harming the disenfranchised rural and urban poor.

In his insightful commentary on veterinary public health in the age of One Health, Primo Arambulo states that it is a challenge to the veterinary profession 'to move the concept of one health forward' (2). But how can this concept ever move forward, and the praxis of One Health be implemented when, as he observes, 'larger and more concentrated units' of animal production are being established around the world to meet 'the rising demand for foods from animal sources'? It would seem that the food animal sector of the veterinary profession, and the aligned

FIGURE 7.1 The mass marketing of pork, beef, fish, and poultry is a major contributor to climate change, loss of natural biodiversity, and consumer health issues.

FIGURE 7.2 This marketing sign of a multinational fast-food chain is a symbol of the conspicuous consumption of 'convenience' and fast foods that are unhealthful dietary choices.

pharmaceutical and animal vaccine industries, in seeking to help meet this rising public demand are at odds with the very principles of the One Health movement The principles and practice of veterinary public health, as articulated by the late Prof. Calvin W. Schwabe (3), surely cannot neglect or deny the environmental,

FIGURE 7.3 Attractive packaging belies the suffering and environmental impact of meat-counter offerings, rarely indicating country of origin and organic, antibiotic, and hormone-free certification.

ecological, economic, and public health costs (especially from food-borne illnesses and zoonoses) that the Western diet, high in animal fat and protein, has brought to those countries where there are concentrated animal feeding operations (CAFOs) and 'affordable' foods from animal sources. CAFOs, coupled with extensive grazing and pastoralist farmed animal production systems, are a major factor in climate change and loss of biodiversity (4).

I am confident, beyond the service role of the veterinary profession to the animal industries, that we have sufficient collective science-based evidence to enlighten and motivate legislators, government policy makers, consumers, health-care providers, and those involved in food animal production. Such a collective involvement in the realities of informed dietary choices and of how animals are raised and treated and how these concerns relate to environmental and public health, as well as to wildlife and biodiversity protection and conservation, is long overdue. Then international efforts, so urgently needed to help steer the rising public demand for foods from animal sources toward those that are only from more humane, certified ecologically sustainable, and ideally organic and healthful systems of animal production, may bear fruit. This is also an essential aspect of achieving sustainable agricultural systems to feed a rising populace (5). It is for our own health and for the good of the Earth, especially considering our numbers, that farmed animal produce can no longer be regarded as dietary staples (6).

FIGURE 7.4 Popular barbequing of meat triggers the Maillard reaction, creating carcinogens in the fat and flesh.

The role of the veterinary profession in helping reduce tropical diseases and zoonoses in developing countries (7), especially through vaccinations, insecticides, parasiticides, and improved monitoring and surveillance, could be counterproductive without supportive family planning and livestock improvements to prevent the harmful consequences of overpopulation and misguided altruism (8).

Consumers and farming organizations in the United States and other countries have accomplished some improved food labeling, including organic certification, genetically modified organism (GMO)–free, and animal welfare certified (Certified GAP.org). But when I worked on the U.S. organic standards with livestock and poultry producers to provide access to the outdoors and recognize animals' basic freedoms, these considerations were rejected in the final government organic standards for farmed animals.

I feel that further documentation of the looming global crisis we all face as a species and as nation-states – East and West – and the many solutions that could be implemented, is called for. For instance, Foley et al. (9) combined information gathered from crop records and satellite images from around the world to create new models of agricultural systems and their environmental impacts that are global in scope. Agriculture now takes up 38% of planet Earth's surface,

FIGURE 7.5 Meatland: A place to shop and drop. This grocery store name is indicative of the world we inhabit where our reliance on cheap meat means 1 billion people are starving or malnourished while another billion are overweight or obese.

with 70% of the wild grasslands, 45% of deciduous forests, and 27% of tropical forests having been converted to agriculture. A third of the world's cropland is dedicated to growing feed for animals and to biofuels. An astounding 75% of the world's entire agricultural land is devoted to growing feed for animals. Whereas approximately 1 billion people are starving or malnourished, another billion are overweight or obese. With a rising population anticipated to reach 9–10 billion in the coming four to five decades, this rate of destruction cannot continue. Some of the key recommendations to double the world's food production while reducing the environmental harms of agriculture in a challenging report by an international team of 20 scientists on how to feed people without ruining the planet include:

1. *Halting farmland expansion and clearing land for agricultural purposes*, particularly in the tropical rainforest. This can be achieved using incentives such as payment for ecosystem services, certification, and ecotourism. This change will yield huge environmental benefits without dramatically cutting into agricultural production or economic well-being.

2. *Shifting diets*: Growing animal feed or biofuels on prime croplands, no matter how efficiently, is a drain on the human food supply. Dedicating croplands to direct human food production could boost calories produced per person by nearly 50%. Even shifting non-food uses such as animal feed or biofuel production away from prime cropland could make a big difference.

3. *Reducing waste*: One-third of the food produced by farms ends up discarded, spoiled, or eaten by pests. Eliminating waste in the path that food takes from farm to mouth could boost food available for consumption by another 50%.

'For the first time, we have shown that it is possible to both feed a hungry world and protect a threatened planet,' said lead author Jonathan Foley, head of the University of Minnesota's Institute on the Environment. 'It will take serious work. But we can do it.'

An earlier report by the FAO (4) documents the global impact of our collective appetite for meat and other farmed animal products as the most damaging of all human industries on Earth. Some of the figures from this report are indeed stunning:

- Livestock account for 20% of the total terrestrial animal biomass.

- Twenty-six percent of the Earth's terrestrial surface is used for livestock grazing.

- Seventy percent of grazing land in dry regions is degraded.

- One-third of all arable land is used to grow livestock feed.
- Eight percent of the world's potable water is consumed by livestock.
- Eighteen percent of greenhouse gas emissions come from livestock.
- Two-thirds of all anthropogenic ammonia, which acidifies rain, comes from the livestock industry.
- In the U.S., the livestock industry is responsible for one-third of all nitrogen and phosphorus loads into the environment.
- The U.S. livestock industry accounts for 37% of all pesticide use and 50% of all antibiotic use.

Regardless of concerted efforts to reduce antibiotic usage in food animals because of the evolution of bacterial resistance – now a worldwide public and animal health problem – the sales of antibiotics intended for domestic food animals increased from 2009 to 2010, according to the U.S. Food and Drug Administration (FDA). The Pew Health Group analyzed the numbers in the report, which showed a boost of 6.7%, from 28.8 million pounds in 2009 to 30.6 million pounds in 2010. If ionophores, used exclusively on animals, are excluded from the analysis, the increase is 8.6%. The increase in antimicrobial sales was greater than the 1.3% increase in meat production, which was up by 1.2 billion pounds to 92.1 billion pounds (see *Food Safety News*, November 1, 2011). This is attributable, in part, to increased exports of pork to China. A new study by a former U.S. FDA economist estimates the total economic impact of food-borne illness across the nation to be a combined $152 billion annually (to obtain a copy of the report, see *Health-Related Costs from Foodborne Illness in the United States, an Initiative of the Pew Charitable Trusts*: visit www.MakeOurFoodSafe.org). It should also be noted that vegetables fertilized with manure from farmed animals fed antibiotics absorb some of these drugs; the risk to consumers is currently being evaluated. The recent call by the European Union to ban antibiotics in animal feed is likely to evoke resistance from the industry and concerted efforts in the United States to promote this enlightened initiative as another fabricated trade barrier, as per the earlier European refusal to accept imported beef from hormone-implanted and anabolic steroid–injected cattle and dairy products from genetically engineered bovine growth hormone–stimulated American dairy cows.

This FAO report (4) recommends a range of measures to mitigate livestock's threats to the environment:

Land degradation: Restore damaged land through soil conservation, silvopastoralism, better management of grazing systems, and protection of sensitive areas.

Greenhouse gas emissions: Sustainable intensification of livestock and feed crop production to reduce carbon dioxide emissions from deforestation and pasture degradation and improved animal nutrition and manure management to cut methane and nitrogen emissions.

Water pollution: Better management of animal waste in industrial production units, better diets to improve nutrient absorption, improved manure management, and better use of processed manure on croplands.

Biodiversity loss: As well as implementing the measures noted earlier, improve protection of wild areas, maintain connectivity among protected areas, and integrate livestock production and producers into landscape management.

This report notes that removal of livestock production subsidies is also likely to improve technical efficiency. In New Zealand, a drastic reduction in agricultural subsidies during the 1980s helped create one of the world's most efficient and environmentally friendly ruminant livestock industries.

According to a Worldwatch Institute report, as part of their ongoing *Nourishing the Planet Project for Vital Signs Online*, global meat consumption and production continue to rise. The institute posted the following summary of their global assessment of meat consumption.

Large-scale meat production also has serious implications for the world's climate. Animal waste releases methane and nitrous oxide, greenhouse gases that are 25 and 300 times more potent than carbon dioxide, respectively.

Dirty, crowded conditions on factory farms can propagate sickness and disease among animals, including swine influenza (H1N1), avian influenza (H5N1), foot-and-mouth disease, and mad-cow disease (bovine spongiform encephalopathy). These diseases not only translate into enormous economic losses each year – the United Kingdom alone spent 18–25 billion dollars in a 3-year period to combat foot-and-mouth disease – but they also lead to human infections.

The amount of meat in people's diets has an impact on human health as well. Eaten in moderation, meat is a good source of protein and of important vitamins and nutrients such as iron, zinc, and vitamins B_3, B_6, and B_{12}. But a diet high in red and processed meats can lead to a host of health problems, including obesity, diabetes, cardiovascular disease, and cancer.

Eating organic, pasture-raised livestock can alleviate chronic health problems and improve the environment. Grass-fed beef contains less fat and more nutrients than its factory-farmed counterpart and reduces the risk of disease and exposure to toxic chemicals. Well-managed pasture systems can improve carbon sequestration, reducing the impact of livestock on the planet. And the use of fewer energy-intensive inputs conserves soil, reduces pollution and erosion, and preserves biodiversity.

Worldwatch Institute

Further highlights from this ongoing Worldwatch Institute research include:

- Pork is the most widely consumed meat in the world, followed by poultry, beef, and mutton.
- Poultry production is the fastest-growing meat sector, increasing 4.7% in 2010 to 98 million tons.
- Worldwide, per-capita meat consumption increased from 41.3 kilograms in 2009 to 41.9 kilograms in 2010. People in the developing world eat 32 kilograms of meat a year, on average, compared to 80 kilograms per person in the industrial world.
- Of the 880 million rural poor people living on less than $1 per day, 70% are partially or completely dependent on livestock for their livelihoods and food security.
- Demand for livestock products will nearly double in sub-Saharan Africa and South Asia, from 200 kilocalories per person per day in 2000 to approximately 400 kilocalories in 2050.
- Raising livestock accounts for roughly 23% of all global water use in agriculture, equivalent to 1.15 liters of water per person per day.
- Livestock account for an estimated 18% of human-caused greenhouse gas emissions, producing 40% of the world's methane and 65% of the world's nitrous oxide.
- Seventy-five percent of the antibiotics used on livestock are not absorbed by the animals and are excreted in waste, posing a serious risk to public health.
- An estimated 11% of deaths in men and 16% of deaths in women could be prevented if people decreased their red meat consumption to the level of the group that ate the least.
- Eating organic, pasture-raised animals can be healthier and more environmentally beneficial compared to industrial feedlot systems.

I am disturbed by the little attention given in any of these reports on agricultural reform and food security to the issue of how farmed animals are transported, handled, and slaughtered. The combination of animal stress and unsanitary conditions increases consumer risk of bacterial food poisoning. As I have witnessed in East Africa and India, improvements in slaughtering operations are urgently needed in developing countries. In industrial countries such as the United States, decentralization to reduce animal transportation distances is a critical animal welfare and public health issue, and the transportation of live animals by sea, as from Australia to Arab states for ritual slaughter, should be prohibited.

The construction of livestock fences for the heavily subsidized beef industry, especially in Botswana, ostensibly to limit diseases that cattle can get from wildlife, has had devastating impact on wildlife unable to reach fresh grazing areas and water by blocking their seasonal migrations (10,11).

The need to adopt organic farming, perennial grains, and humane, sustainable livestock feeding methods – especially grass-fed and mixed-crop systems – have been underscored by the U.S. National Research Council (12) addressing what is needed to transform U.S. agriculture. Economic triage calls for an end to price supports and subsidies currently going to the livestock industry and allied animal feed industry that promote non-sustainable commodity crop production (particularly genetically engineered corn and soy), with harmful consequences for consumers and the environment. These consequences are especially severed from agrichemical pollution and soil runoff, causing sedimentation of lakes and rivers and the growing 'dead zone' in the Gulf of Mexico.

The massive assault by the livestock industry on the now globally threatened marine ecosystems must also be addressed, and not only from nutrient waste runoff and agrichemicals in eroded soils from arable land used to produce livestock feed. A vast tonnage of krill and fish 'bycatch' occurs from indiscriminate overfishing and creates profit for the fishing industry, which incorporates the bycatch into livestock feed. This is yet another hidden cost of the animal farming industry that includes farmed salmon and shrimp.

I feel somewhat vindicated by these science-based reports documenting the problems created by industrial agriculture after being a relatively lone voice over two decades ago (13) on this complex issue, and more recently over the ethics and consequences of genetic engineering agribiotechnology (14). It is now clearly documented that genetically engineered crops have helped promote the evolution of herbicide-resistant 'superweeds,' helped undermine food security in developing countries, resulted in increased pesticide use, and failed to increase the yield of any human food crop (15). But scientific documentation and economic evidence alone will not stop this global food industry juggernaut and bring the long-overdue changes to improve consumer safety, food security, environmental protection, production efficiencies, and farmed animal health and welfare. We need to make the entire agricultural enterprise sustainable and socially just. In the face of political inertia, multinational corporate collusion and corruption, and a monopolistic monolith of agribusiness with its pharmaceutical and petrochemical industry allies and lobbyists, a food revolution is nonetheless possible. It is beginning now as informed consumers vote with their dollars for food labeling, purchase locally grown produce, learn how to prepare nutritious meals from basic ingredients, and favor organically certified and non-GMO-verified produce.

Biologist Prof. Rene Dubos (16) forewarned that 'A relationship to the earth based only on its use for economic enrichment is bound to result not only in its degradation but also in the devaluation of human life. This is a perversion which, if not corrected, will become a fatal disease of technological societies.' Millions of people today are suffering the consequences of what Prof. Dubos predicted, as others suffer the burdens of empathy for others and animals, wild and domestic, are caught up in the current downward spiral of the global economy and an uncertain future for all. The cold realities of human triage in process are now more through default than by ethical decision, a fact some see as a conspiracy of the rich over the poor and which Mahatma Gandhi encapsulated in his quip that 'The cattle of the rich steal the bread of the poor.'

One of America's first environmentalists, Aldo Leopold (17), wrote, 'The cowman who clears the range of wolves does not realize that he is taking over the wolf's job of trimming the herd to fit the range. He has not learned to think like a mountain. Hence, we have dust bowls and rivers washing the future into the sea.' The continuing government subsidies provided to cattle ranchers grazing public lands and for indiscriminate and cruel predator control speak to the power of the livestock lobby for whom sustainability is anathema.

Now it is time for governments to break from the yoke of corporate hegemony and acknowledge the new economic reality the world faces today (18) and to withdraw support with public taxes and subsidies for harmful agricultural practices, along with informed consumers voting with their dollars. These measures will expedite the transformation of all industries and their products and services into ethical and sustainable human activities. This should begin in the agriculture/food and fuel/energy sectors, where the public still has considerable influence, along with allied banks, lending institutions, universities, and transnational aid and development agencies governed by the global bioethics (19) of social and eco-justice and sustainability.

A full accounting of the externalities or hidden costs of conventional industrial agriculture and, in particular, its commodity crop and livestock sectors, would soon put an end to the largess of government price supports and subsidies and discourage banks and international aid and development agencies from providing loans to non-sustainable and environmentally harmful farming and other enterprises. The public health costs of the modern Western diet, with its associated cancers, obesity epidemic/metabolic syndrome, and other diet-related diseases and food-borne illnesses, must also be considered, causing organic farming practices, vegetarianism, veganism, and 'conscientious omnivorism' (consuming less, and choosing only humanely and organically produced foods of animal origin) to be more widely appreciated as ethical imperatives and enlightened self-interest. The health benefits of the so-called Mediterranean diet (which is also of environmental and animal welfare benefit) that focuses on plant-based foods (whole grains, fruits, nuts, vegetables, and extra-virgin olive oil), with minimal consumption of fish, meat, and dairy products, have been closely documented (20). Women who closely followed the Mediterranean diet had a 23% lower overall mortality risk during the period of study, with a higher prevalence of cardiovascular disease and cancer in women in the study who did not adhere to the proposed diet (21). I would call this a biologically appropriate diet, rich in prebiotics, along with the addition of probiotic-rich fermented foods such as sauerkraut and kimchi for *Homo sapiens* (which is healthful for dogs but not for cats!). Good nutrition supports a strong immune system and a diverse gut microbiome, including some species of bacteria that produce precursors of the 'bonding' and metabolism-regulating hormone, oxytocin (22). Higher oxytocin levels in consumers would increase an overall sense of well-being and decreasing depression and aggression in the populace.

FIGURE 7.6 More and more studies are confirming the health, economic, and environmental benefits of plant-based diets.

Industrialized nations need to provide more in-field veterinary services to help improve livestock health and to prevent the spread of 'transboundary' diseases in developing countries since, according to veterinarian Dr. Peter Roeder (23), 'There is an intimate relationship between "homeland security" and international health and development issues.' Dr. Roeder played a key role in the final stages of eliminating rinderpest (cattle plague) from the world that has devastated livestock and wildlife for centuries. He documents how more intensive, industrial-scale livestock and poultry production systems, currently expanding in developing countries to meet the increasing demand for meat, milk, and other animal foods, are contributing to the emergence and spread of epidemic diseases at an increasing rate, noting that:

- Infectious bovine rhinotracheitis, once a sexually transmitted infection, evolved in feedlot beef cattle in the United States to become a severe respiratory disease.
- Intensification of goat production in the Netherlands helped make Q fever a major health issue.
- Intensification of duck production in China and Southeast Asia set the conditions for the avian flu virus H5N1HPAI to become a global pandemic.
- Intensification of dairy production in Saudi Arabia created difficulties in controlling foot and mouth disease.
- Multiple factors associated with intensive production systems contribute to immune system impairment and increased susceptibility to disease in poultry, beef and dairy cattle, and pigs. We can also add fish in aquaculture.

Waste is sinful, even in times of plenty. Food waste in developed countries from marketing and personal household practices is indeed scandalous. The amount exceeds 8 million tons in the UK and 40 million tons in the United States annually (24). Much of this waste, still suitable for human consumption, could, with appropriate management and processing, be profitably processed into feed for poultry and pigs along the lines of model legislation in Japan, which directs local authorities to recycle a portion of waste food into animal feed.

Over 50 years ago, I sought to draw government attention, along with other concerned scientists, physicians, and veterinarians, about the inclusion of antibiotics in farmed animals' feed. Antibiotics were used liberally to boost productivity and prevent the bacterial disease to which animals were more susceptible when raised in overcrowded factory farms and feedlots – spreading globally under the delusion of progress – often funded by the World Bank and international aid and development organizations. There was increasing evidence of harmless bacteria developing resistance to antibiotics and even passing on this resistance to more harmful bacteria, thus reducing the effectiveness of such drugs when called for to treat sick animals and people. Today we are paying the price. More than 1.2 million people died in 2019 of infections from antibiotic-resistant bacteria, according to a global study of 204 countries. The burden was highest in sub-Saharan Africa and South Asia, with children under 5 the most at risk. Another 5 million people died in 2019 from diseases in which antibiotic-resistant bacteria had a role. For comparison, researchers estimate that, in the same year, AIDS caused 860,000 deaths and malaria caused 640,000 (25).

It is time to prohibit the pharmaceutical industry's wholesale global marketing of all production-enhancing drugs to farmed animals, because of various adverse effects on animals' welfare and the environment, as well as consumer risks.

PLANT-BASED DIETS TO HELP RESTORE BIODIVERSITY AND REDUCE EMISSIONS

The prestigious medical journal the *Lancet* has taken a leading role in promoting dietary choices that are better for our own health and for the environment and other species – wild and domesticated – with whom we share this planet and which we are now rapidly destroying. Many people are now adopting the EAT-Lancet diet (26).

Scientists have now determined that if high-income countries were to adopt this diet, the energy savings and carbon sequestering would help significantly reduce climate change and loss of biodiversity. Adopting a more plant-based diet could give rich countries a 'double climate dividend' of lower emissions and more land for capturing carbon (27). There is also a health dividend: Plant-based and mainly vegetarian diets were associated with a lower incidence of COVID-19 infection and may even be considered protective against it (28). It is no conspiracy theory to propose or suppose that this pandemic was one of nature's correctives since this planet cannot sustain 8 billion human omnivores and adopters of the 'paleo' meat-based diet without further exacerbating climate change and the displacement and extinction of other species.

It is also notable and encouraging that the practice of recycling farmed animal remains, including parts condemned for human consumption, into pet foods is being confronted by increasing numbers of pet food manufacturers. Plant-based and alternative protein and fat ingredients (including farmed insects) are being developed with the support of sound veterinary evidence of health benefits for companion animals (29).

The nascent industry of producing meat and fish from animals' cells through bio-fermentation also holds promise, eliminating animal suffering and the negative environmental impact of CAFOs. But there is already political opposition in step with the U.S. Supreme Court ruling against the State of California's public support of Proposition 2. Enacted January 1, 2015, it prohibits, with certain exceptions, the confinement on a farm of pregnant pigs, calves raised for veal, and egg-laying hens in a manner that does not allow them to turn around freely, lie down, stand up, and fully extend their limbs, along with a prohibition of importation of animal produce from other states not measuring up to these more humane standards. Now, in June 2024, a U.S. Republican congressman proposed a bill that would ban the use of federal funds in various areas related to cultured meat. Eight Republican colleagues cosponsored Representative Warren Davidson's (R-OH) REAL Meat Act of 2024. 'Fake cell-cultured meat not only poses a health risk to the human body, but it also threatens the livelihoods of America's hard-working ranchers, livestock farmers, and butchers,' Davidson said after the bill was introduced. 'Recently, laws prohibiting lab-grown meat have been passed in states like Florida and Alabama. Congress must act to ensure US taxpayers are not footing the bill for this inferior, experimental product.'

I know of no documentation of any such health risks, but they are well documented with conventional animal products. Cultured meats and fish could provide greater consumer safety and significantly less negative environmental impact and contribution to climate change – and animal suffering. Perhaps these Republican shills of the livestock industry will seek to block a UK company that plans to bring the first cultivated-meat cat food to the United States through an R&D partnership with a manufacturer in the that country, where regulatory tests also remain, says Owen Ensor, co-founder and CEO of Meatly, which sells 5-ounce cans in the UK for about $1.25. Meatly's declared mission is to end pet food's reliance on farmed animals and decrease its effect on the environment.

Since the *Lancet* report advocating plant-based diets (30), a new review emphasizes cultural and bioregional considerations of acceptability and sustainability (31). Advocating biologically and bioregionally appropriate diets for our species may help prevent many diseases. A study in mice and people suggests why there is a link between obesity and some cancers: a high-fat diet increases the number of *Desulfovibrio* bacteria in the gut (32). These release leucine, an amino acid, which encourages the proliferation of a kind of cell that suppresses the immune system. With a suppressed immune system, breast cancer tumor growth increases.

I would be remiss in not underscoring the fact that farmed salmon in particular is not a "health food." Emamectin is fed to caged salmon, stressed with overcrowding confinement, to kill the lice that can plague them. But it is excreted and pollutes lochs around Scotland, killing prawns, crabs, lobsters, and other crustaceans. Tests of farmed salmon from U.S. grocery stores show that farmed salmon are likely the most PCB-contaminated protein source in the U.S. food supply. On average farmed salmon have 16 times the dioxin-like PCBs found in wild salmon, 4 times the levels in beef, and 3.4 times the dioxin-like PCBs found in other seafood (33). Salmon accumulate PCBs from the fishmeal they are fed, so-called 'bycatch,' contributing to overfishing. Scottish farmed salmon is so contaminated with PCBs, dioxins, dieldrin and toxaphene that no more than three meals of Scottish farmed salmon per year are recommended (34). The addition of discarded salmon parts to pet foods and salmon oil supplements needs to be questioned.

PFAS ("forever" chemicals) and microplastics are also a consumer issue in salmon and other seafoods. Plastic pollution is creating a 'plastisphere,' a widespread habitat that includes pathogenic viruses and antimicrobial-resistant bacteria that affect the health of plants, animals, and humans. *Vibrio* bacteria, for example, which are normally rare in the open ocean, are widely distributed in plastispheres throughout the mid-North Atlantic Ocean where they can cause diseases in marine life, including fish, shellfish, and coral, as well as in humans (35). Perhaps it is time for this sector of aquaculture to close or transition to sustainable, carbon-sequestering kelp farming.

The health benefits of humans and companion animals consuming some crustaceans, insects and mushrooms high in chitin are implicated in research demonstrating immune system enhancement (36).

Persistent Organic Pollutants (POPs) including PCBs, dioxins, furans, brominated flame retardants, and perfluorinated compounds (PFCs) are lipophilic and accumulate in biota, meaning that consumers are at risk through ingestion of foods from higher levels of the food chain, like meat, dairy products, fish, and shellfish. In utero exposure and exposure due to ingestion of breast milk also occur and are exposure pathways of concern (37). Since POPs accumulate up the food chain, are linked to many health issues including neurodegenerative diseases and cancer, this is another reason why many informed consumers are eating lower on the food chain by opting for vegan and transitional vegetarian diets, and some veterinarians are now advising similar dietary changes for companion animals.

Reducing the farmed animal population and wildlife traffic and trade for human consumption would also help prevent zoonoses and pandemics, which will be discussed in the next chapter.

REFERENCES

1. FAO report, State of the World's Land and Water Resources for Food and Agriculture. https://www.fao.org/land-water/land/en/
2. Arambulo, P. 3rd. Veterinary public health in the age of 'one health'. *Journal of the American Veterinary Medical Association*, 2011; 239: 48–49.
3. Schwabe, CW. *Veterinary Medicine and Human Health*. 3rd ed. Baltimore, MD. Williams & Wilkins, 1984.
4. Steinfeld, H.P. et al. *Livestock's Long Shadow: Environmental Issues and Options*. Washington, DC, United Nations Food and Agriculture Organization, 2006.

5. Reganold, J.P., et al. Transforming U.S. agriculture. *Science*, 2011; 332: 670–671.
6. Hu, F.B. and Willett, W.C. The relationship between consumption of animal products (beef, pork, poultry, eggs, fish, and dairy products) and risk of chronic disease: A critical review. Report for the World Bank. Cambridge, MA, Harvard School of Public Health, 1998. See also: Campbell, T.C. *The China Study: The Most Comprehensive Study of Nutrition Conducted, and the Startling Implications for Diet, Weight-loss and Long-term Health*. Dallas, TX, Bella Books, 2005.
7. Welburn, S. One Health: The 21st century challenge. *Veterinary Record*, 2011; 169: 614–615.
8. Hardin, G. *The Limits of Altruism: An Ecologist's View of Survival*. Bloomington, IN, Indiana University Press, 1997.
9. Foley, A.J. et al. Solutions for a cultivated planet, *Nature*, 2011; 478: 337–342.
10. Owens, M. and Owens, D. The fences of death. *African Wildlife*, 1993; 34: 25–77.
11. Mbaiwa, J.E. and Mbaiwa, O.J. The effects of veterinary fences on wildlife populations in Okavango Delta, Botswana. *International Journal of Wilderness*, 2006; 12: 17–23 & 41.
12. National Research Council. *Toward Sustainable Agricultural Systems in the 21st Century*. Washington, DC, National Academies, 2010.
13. Fox, M.W. *Agricide: The Hidden Crisis That Affects Us All*. New York, Schocken Books, 1988.
14. Fox, M.W. *Superpigs and Wondercorn: The Brave New World of Biotechnology – and Where It All May Lead*. Guilford, CT, Lyons Press, 1992.
15. Global Citizens' report on the State of GMOs. The GMO emperor has no clothes: False promises, failed hopes, 2011. Available at www.navdanya.org
16. Dubos, R. *A God Within*. New York, Scribners, 1972.
17. Leopold, A. *A Sand County Almanac*. New York, Oxford University Press, 1949.
18. Heinberg, R. (ed). *The End of Growth: Adapting to Our New Economic Reality*. Gabriola Island, British Columbia, Canada, New Society Publishers, 2011.
19. Fox, M.W. *Bringing Life to Ethics: Global Bioethics for a Humane Society*. Albany, NY, State University of New York Press, 2001.
20. Soltani S. et al. Adherence to the Mediterranean diet in relation to all-cause mortality: A systematic review and dose-response meta-analysis of prospective cohort studies. *Advances in Nutrition*, 2019; 10(6): 1029–1039.
21. Ahmad S. et al. Mediterranean diet adherence and risk of all-cause mortality in women. *JAMA Network Open*, 2024 May 1; 7(5): e2414322.
22. Erdman, S.E. and Poutahidis, T. Microbes and oxytocin: Benefits for host physiology and behavior. *International Review of Neurobiology*, 2016; 131: 91–126. See also, Varian, B.J. et al. Beneficial dog bacteria up-regulate oxytocin and lower risk of obesity. *Journal of Probiotics & Health*, 2016; 4: 1–9.
23. Roeder, P. Making a global impact: Challenges for the future. *Veterinary Record*, 2021; 169: 671–764.
24. Stuart, T. *Waste: Uncovering the Global Food Scandal*. London, Penguin Books, 2009.
25. Antimicrobial Resistance Collaborators. Global burden of bacterial antimicrobial resistance in 2019: A systematic analysis. *Lancet*, 2022; 399(10325): 629–655. https://doi.org/10.1016/S0140-6736(21)02724-0
26. For the detailed report go to https://eatforum.org/learn-and-discover/the-planetary-health-diet/ and https://eatforum.org/eat-lancet-commission/eat-lancet-commission-summary-report/
27. Zhongxiao, S. et al. Dietary change in high-income nations alone can lead to substantial double climate dividend. *Nature Food*, 2022; 3: 29–37. https://doi.org/10.1038/s43016-021-00431-5

28. Acosta-Navarro, J.C. et al. Vegetarian and plant-based diets associated with lower incidence of COVID-19. *BMJ Nutrition, Prevention & Health*, 2024; e000629. https://doi.org/10.1136/bmjnph-2023-000629
29. Knight, A. The relative benefits for environmental sustainability of vegan diets for dogs, cats and people. *PLOS ONE*, 2023; 18(10): e0291791. https://doi.org/10.1371/journal.pone.0291791
30. Willett, W. et al. Food in the Anthropocene: The EAT–Lancet Commission on healthy diets from sustainable food systems. *Lancet*, 2019; 393: 447–492
31. Loken, B., et al. Healthy and sustainable diets must be culturally acceptable too. *Nature Food*. 2024; Aug 30: 1–2.
32. Chen, J. et al. A high-fat diet promotes cancer progression by inducing gut microbiota–mediated leucine production and PMN-MDSC differentiation. *PNAS*. May 6, 2024; 121(20): e2306776121. https://doi.org/10.1073/pnas.2306776121
33. https://www.ewg.org/research/pcbs-farmed-salmon (Accessed October 12, 2024).
34. Hites, R.A.et al. Global assessment of organic contaminants in farmed salmon. *Science*. 2004; 303: 226–229.
35. Bowley, J. et al. Oceanic hitchhikers – Assessing pathogen risks from marine microplastic. *Trends in Microbiology*, 2021; 29: 107–116.
36. Do-Hyun, K. et al. A type 2 immune circuit in the stomach controls mammalian adaptation to dietary chitin. *Science*, 2023; 381: 1092–1098.
37. https://www.epa.gov/expobox/exposure-assessment-tools-chemical-classes-other-organics (Accessed November 1, 2024).

8 COVID-19 and Preventing Other Animal-to-Human Diseases

This chapter considers the genesis of zoonotic diseases like the recent global COVID-19 pandemic as they related to our treatment and consumption of other species. How we respond to such crises, their emotional and socioeconomic consequences, and what changes in human behavior are called for to prevent such catastrophic, anthropogenic health crises more effectively is well encapsulated in these words by two physicians:

> The time has come for us to rethink our relationship with all life on this planet – other humans, nonhumans, and planet Earth, a life form in itself: Gaia. What is good for nonhumans and the Earth is virtually always in the best interests of humans, given the profound interconnectedness of all life. All that we do depends upon abundant plant and animal life as well as clean air and water. Each of us can have a positive impact upon these fundamental resources by demonstrating and inspiring an enhanced mindfulness, beginning most basically with what we eat and how all our daily choices and actions may be affecting animals and natural habitats. Ultimately, the survival of not only other life forms on this planet, but also our own, will depend upon humanity's ability to recognize the oneness of all that exists and the importance and deeper significance of compassion for all life. (1)

This devastating pandemic and other transmissible diseases that are likely to become pandemics in the future call for ever more vaccines and medications. These are not risk-free, but will be our main costly defense, so long as preventive medicine remains human-centered and does not address, under the banner of One Health, our relationships with and treatment of other species. Specific issues are wildlife farming, poaching, and trafficking; habitat encroachment; our ever-increasing human numbers; and consumption of animals, wild and domesticated.

The concept of One Health is not new, and perhaps has even enjoyed stronger endorsement and support in past decades. prior to the advent of clinical specialization in human and veterinary medicine. Steps to achieving the end point of this concept and seeing it put into action internationally are well articulated by the One Health Initiative (2). The COVID-19 pandemic, with more pandemics predicted, puts the applicability concept and its activation in bold relief.

DOI: 10.1201/9781003569787-8

A rise in zoonotic diseases is being driven by environmental degradation, according to a report by the UN Environment Program and the International Livestock Research Institute, which cites rising demand for animal protein, intensive farming practices, exploitation of wildlife, and climate change, among other key factors. The authors suggest adopting a One Health approach, which would unite public health with veterinary and environmental experts to respond to and prevent zoonotic disease outbreaks (3).

Giving the public hope via protective, animal-tested vaccinations now being developed around the world, and evaluating various drugs to treat infected patients may be to little avail, considering how pathogens like the SARS-CoV-2 virus can mutate into a new strain or variant causing a different set of health problems and various symptoms according to age, pre-existing health issues, sex, and race and infecting other species, including deer and cats. Vaccination limitations are a documented problem with the influenza virus, meaning some vaccine formulations carry intrinsic vaccinosis risks and do not provide adequate protection. Also, some vaccinations can lead to increased susceptibility to other viral infections.

Regrettably, organized veterinary medicine in the farm/food animal sector has prioritized human interests of profitability and productivity over animal health and well-being. In poor countries and communities, animal health needs of small producers are not adequately served, and corruption, falsifying vaccination records, and inadequate surveillance and prevention of zoonotic diseases have been well documented (4).

Millions of mammals, amphibians, birds, insects, and reptiles are imported legally into the United States every year, potentially bringing with them 'a kaleidoscope of pathogens,' writes former Fish and Wildlife Service inspector Jonathan Kolby. 'With few exceptions, the US has no laws specifically requiring disease surveillance for wildlife entering the country, and the vast majority of wild animal imports are therefore not tested,' Kolby writes (5).

While this reality may make one despair, all countries should be severely sanctioned economically for engaging in wildlife trafficking and for having open 'wet markets' selling wild-caught animals. And there must be a redoubling of wild habitat protection from human encroachment with population control through voluntary and ready access to family planning, smaller families, and communities needing fewer livestock to sustain their needs.

Philosopher David Benatar observes, 'It is curious, therefore, that changing the way humans treat animals – most basically, ceasing to eat them or, at the very least, radically limiting the quantity of them that are eaten – is largely off the radar as a significant preventive measure' (6).

An article in the *Intergovernmental Science-Policy Platform on Biodiversity and Ecosystem Services* puts the responsibility for COVID-19 squarely on our shoulders: 'There is a single species that is responsible for the COVID-19 pandemic – us. As with the climate and biodiversity crises, recent pandemics are a direct consequence of human activity – particularly our global financial and economic systems, based on a limited paradigm that prizes economic growth at any cost. We have a small window of opportunity, in overcoming the challenges of the current crisis, to avoid sowing the seeds of future ones' (7).

One hopes that this global health crisis will catalyze international collaboration in prevention beyond reliance on more vaccines and pharmaceutical products. Above all, humans should keep out of wildlife habitat where such diseases emerge and to which we have no immunity. Ideally, consumers in industrial countries will support producers of organically certified foods to sustain healthful vegetarian and vegan diets with minimal or zero consumption of eggs, dairy, or meat, including seafood.

Continuing to consume animals as a basic food source, marketing ever more vaccines, and having ever more children, the rich and poor alike will be subject to the indiscriminate justice of natural law until we all live in greater harmony with other species as well as with each other. Alternatively, with deteriorating natural controls of health-sustaining biodiversity, plagues and pestilences of biblical proportions will be the legacy of our collective failure in planetary stewardship. Preventing pandemics and other animal/food-borne epidemics and regional outbreaks of disease will be impossible without limiting factory farm animal production systems – billions of poultry and pigs worldwide that are the primary source of various strains of influenza virus and antibiotic-resistant strains of bacteria. It is a matter of public health and animal health and well-being to end all intensive forms of food animal production.

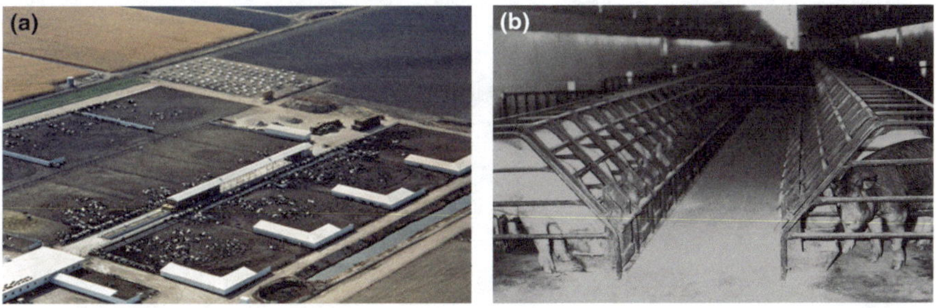

FIGURE 8.1 (a) A U.S. hog factory farm with thousands of confined animals that suffer extreme weather events and create a waste management issue often responsible for ground and surface water pollution. (b) Sows kept in severely restricted stalls. Preventing pandemics and other animal/food-borne epidemics and regional outbreaks of disease will be impossible without limiting factory farm animal production systems like these.

It is also essential to give wildlife space and reduce human contact and conflicts. Ecologist Raina Plowright and associates work with black flying foxes and have demonstrated how preserving their natural habitat can minimize contact with humans and prevent spillover of the Hendra virus, leading to their advocacy for a holistic One Health approach, according to public health researchers, to minimize spillovers and prevent epidemics and pandemics. This multidisciplinary team recommends protecting animal food sources to reduce encounters with humans, safeguarding areas where animals congregate to minimize stress, and protecting the humans who interact with animals with adequate gear and education (8).

Climate change, biodiversity loss, the introduction of non-native species, and chemical pollution substantially contribute to a rise in emerging infectious diseases, according to a meta-analysis of pandemic etiology (9). Biodiversity loss was found to be the biggest driver of infectious disease spread, possibly because pathogens evolve to favor common species, but perhaps paradoxically, urban areas tend to have less disease, partly due to public sanitation and access to health care, but also due to fewer wild animals that transmit diseases. This analysis concludes that 'reducing greenhouse gas emissions, managing ecosystem health, and preventing biological invasions and biodiversity loss could help to reduce the burden of plant, animal and human diseases, especially when coupled with improvements to social and economic determinants of health.' At best, all the suffering, death, grieving, and economic impact of this latest COVID-19 pandemic will change how we choose to live. Ideally it will reduce our collective exploitation and consumption of animals that bring on such pandemics and other zoonotic diseases.

FIGURE 8.2 Crowded egg-laying hens in so-called battery cages with evident lesions from feather picking due to the stress of extreme confinement.

On the horizon at the time of writing is the highly infectious H5N1 avian influenza virus, the ancestral strain of which initially emerged in farmed poultry in Guangdong, China, in 1996 and underwent continuous evolution as it circulated in Africa and Eurasia, with migratory birds bringing it to Europe, the Americas, and other parts of the world. Dairy cattle and cats on dairy farms have become infected, but few human infections have been reported. The first large-scale spillover identified in wild mammals was an outbreak in harbor and grey seals along the coast of Maine. Subsequently, wild marine and terrestrial carnivores, from polar bears to bobcats, have died after consuming infected wild birds. Mink, fox, and raccoon dogs on fur farms in Europe have been infected. Hundreds of millions of intensively farmed poultry have been euthanized (by switching off ventilation) in many countries to prevent the spread of this disease. In July 2024 it was reported that Moderna will use $176 million from the U.S. Department of Health and Human Services to speed development of an mRNA H5N1 influenza vaccine that could be quickly revised if a different form of influenza were to become a larger threat.

In 2011, Fouchier and Kawaoka alarmed the world by revealing they had separately modified the H5N1 influenza virus so that it spread between ferrets. Advocates of such gain-of-function (GOF) studies said they can help public health experts better understand how viruses might spread and plan for pandemics. But by enabling the bird virus to more easily spread among mammals, the experiments also raised fears that the pathogen could jump to humans. Critics of the work worried that

FIGURE 8.3 Standard conditions for broiler chickens raised for their meat are accepted worldwide in total disregard for the birds' well-being and health, turning these systems into Petri dishes for the proliferation of diseases, notably avian influenza.

such a souped-up virus could spark a pandemic if it escaped from a lab or was intentionally released by a bioterrorist. After extensive discussion about whether the two studies should even be published (they ultimately were) and a voluntary moratorium by the two labs, the experiments resumed in 2013 under new U.S. oversight rules (10). But concerns reignited after more papers and a series of accidents at federal biocontainment labs. In October 2014, U.S. officials announced an unprecedented pause on funding for 18 GOF studies involving influenza, Middle East respiratory syndrome, and severe acute respiratory syndrome viruses (11,12).

Pandemic prevention calls for greater vigilance at the human–animal interface where animals become infected by people, creating infection reservoirs in animal populations. While SARS-CoV-2 has a potential broad wildlife host range, most documented infections have been in captive animals and a single wildlife species, the white-tailed deer. In one report, epidemiologists sampled 23 species of wildlife for SARS-CoV-2 and examined the effects of urbanization and human use on seropositivity. SARS-CoV-2 RNA was found in six species, including the deer mouse, Virginia opossum, raccoon, groundhog, Eastern cottontail, and Eastern red bat, between May 2022 and September 2023 across Virginia and Washington, DC. Sites with high human activity had three times higher seroprevalence than low human-use areas. This study shows that exposure to SARS-CoV-2 has been widespread in wildlife communities and suggests that areas with high human activity may serve as points of contact for cross-species transmission (13).

The public health costs and pandemic potential of factory farms are exemplified by the poultry industry. Of the estimated over 1 million cases of food-borne salmonellosis acquired annually in the United States, over 24% are attributed to consumption of chicken and turkey products (14). Despite extensive *Salmonella* controls used at processing, 5.5% of salmonellosis cases are linked to turkey (15). The U.S. Department of Agriculture proposed a rule in July 2024 that poultry operations would be required to test for six specific *Salmonella* strains and limit overall *Salmonella* levels in their products. After a 3-year phase-in period, poultry operations would have to establish monitoring programs and follow guidelines to reduce the risk of *Salmonella* spreading through flocks. Coincidentally at this same time dairy cows and their milk were being tested because many cows succumbed to the H5N1 avian influenza virus, and there were concerns this viral strain might mutate and enter the human population, already mildly infecting a few dairy farm workers and, more fatally, farm cats.

Drastically reducing the burning of fossil fuels not only reduces the severity of climate change but also prevents diseases and increased susceptibility to infection and cancer caused by small particle pollutants (PM2.5), which is an international responsibility for a United Environmental Nations. Fine particulate matter, or PM2.5 (which contains lead, zinc, aluminum, and sulfuric acid), in the air we breathe that comes from traffic, factories, and forest fires can damage our lungs and hearts, cause cancer, and may damage the brain and play a role in dementia. The impact on wildlife, both terrestrial and aquatic, is extensive, affecting their survival behaviors and neuroendocrine, immune, and reproductive systems (16). This widespread pollutant makes lakes and streams acidic, changes the nutrient balance in coastal waters and large river basins, depletes the nutrients in soil, damages sensitive forests and farm crops, and affects the diversity of ecosystems (17).

According to the Louisiana State University School of Veterinary Medicine, outbreaks of canine respiratory disease complex (CRDC) reached a total of 16 states by December 2023: California, Colorado, Florida, Georgia, Idaho, Illinois, Indiana, Maryland, Massachusetts, New Hampshire, Nevada, Oregon, Pennsylvania, Rhode Island, Vermont, and Washington. No single infectious agent has been identified in these not-infrequent local and regional outbreaks of CRDC in dogs. Fine particulate matter that dogs sniff and inhale while panting when outdoors walking and running close to the ground could be a contributing factor, especially since the emergence of CRDC is coupled with ever more frequent and extensive forest fires across Canada and the United States.

Keeping wild animals in cages and raising them for their fur and especially in China for traditional medical products needs to be prohibited. A research team detected 125 viruses, including 36 new ones, according to the study in the journal *Nature*. Thirty-nine of the viruses have a ‘high risk’ of jumping across species, including to humans, the researchers evaluated. Some of those viruses – such as hepatitis E and Japanese encephalitis – have already spread to humans, but 13 were new. Pipistrellus bat HKU5-like virus previously detected in bats was found in the lungs of two farmed minks. It is a relative of the Middle East respiratory syndrome coronavirus (MERS), which can be deadly to humans.

Several types of bird flu were also detected in guinea pigs, minks, and muskrats, and seven types of coronaviruses were also spotted – though none were closely related to SARS-CoV-2, which causes COVID (18).

To close on a lighter note, several years ago, I was attending a conference on the pros and cons of genetic engineering biotechnology in a small and crowded room at the Royal Society in London. In the middle of the meeting, seeing several heads were nodding off, I got up and opened all the windows along one side of the conference room. I felt like one of those proverbial frogs in a container that was being slowly heated and would have gone to sleep, but would have jumped out immediately if the water was already hot! The passive acceptance of the ‘stuffiness’ of that room was evidence of how we tend to adapt to, or accept, conditions potentially detrimental to our health. It is notable that elevated levels of carbon dioxide favor the survivability and transmissibility of viral infections (19). This issue was not fully addressed at the beginning of the COVID-19 pandemic in preventing spread, especially in schools and enclosed public spaces. It is notable that veterinarians have established 10–15 fresh air exchanges for laboratory animal facilities, while schools have much lower rates of air exchange. Also, air quality in terms of ionization can affect health and pathogen survivability (20).

The next chapter considers the role of applied veterinary bioethics in animal and planetary health and what correctives are called for.

NOTE

An article by Lura H. Kahn and Richard Seifman 'It's Time to Get Serious about Preventing Pandemics: A New World Bank/WHO Fund,' suggests that prevention is a priority and that the One Health interdisciplinary approach is critical. The article underscores the need for a more holistic approach by responsible international aid and health agencies. Kahn and Seifman state,

> The good news is that the World Bank with the World Health Organization is about to launch a brand-new international multilateral financing mechanism to provide long-term funding to address pandemics.... Unfortunately, the preparedness and response parts of the mechanism are receiving the lion's share of the attention and funding compared to the prevention part. We must include veterinarians and other animal health professionals who receive extensive education and training in zoonotic diseases and environmental health experts to be part of the pandemic [One Health] prevention solution. So far, their involvement has been minimal which jeopardizes the effort's success. (https://impakter.com/time-get-serious-about-preventing-pandemics/)

REFERENCES

1. Wiebers, D.O. and Feigin, V.L. What the COVID-19 crisis is telling humanity. *Neuroepidemiology*, 2020; 54(4): 283–286.
2. One Health Initiative. The 21st Century US and Worldwide Landmark Catalyst for One Medicine One Health. June 20, 2020. https://onehealthinitiative.com/the-21st-century-u-s-and-worldwide-landmark-catalyst-for-one-medicine-one-health/
3. United Nations News. New UN report outlines ways to curb growing spread of animal-to-human diseases. July 6, 2020. https://news.un.org/en/story/2020/07/1067711
4. Krantz, D.L. and Fox, M.W. *India's Animals: Helping the Sacred and the Suffering*. CreateSpace Books, Amazon.com, 2016.
5. Kolby, J. To prevent the next pandemic, it's the legal wildlife trade we should worry about, *National Geographic*. https://www.nationalgeographic.com/animals/2020/05/to-prevent-next-pandemic-focus-on-legal-wildlife-trade/
6. Benatar, D. Editorial: The chickens come home to roost. *American Journal of Public Health*, 2020; 97: 1546.
7. Settele, J.D. et al. COVID-19 Stimulus Measures Must Save Lives, Protect Livelihoods, and Safeguard Nature to Reduce the Risk of Future Pandemics. April 27, 2020. Intergovernmental Science-Policy Platform on Biodiversity and Ecosystem Services.
8. Plowright, R.K. et al. Ecological countermeasures to prevent pathogen spillover and subsequent pandemics. *Nature Communications*, 2024; 15: 2577. https://doi.org/10.1038/s41467-024-46151-9

9. Maon, M.B. et al. A meta-analysis on global change drivers and the risk of infectious disease. *Nature*, May 2024; 629(8013):830–836.
10. Enserink, M. (2011). Scientists brace for media storm around controversial flu studies. *ScienceInsider*, November 23, 2011. http://news.sciencemag.org/scienceinsider/2011/11/scientists-brace-for-media-storm.html
11. Kaiser, J. and Malakoff, D. U.S. halts funding for new risky virus studies, calls for voluntary moratorium. *Science*, 2014; 346:404.
12. Goldberg, A.R. et al. Widespread exposure to SARS-CoV-2 in wildlife communities. *Nature Communications*, 2024; 15: 6210.
13. Fritz, M. et al. High prevalence of SARS-CoV-2 antibodies in pets from COVID-19+ 13 households. *One Health*, 2020; 11: 100192.
14. Scallan, W.E. et al. Estimating the number of illnesses caused by agents transmitted commonly through food: A scoping review. *Foodborne Pathogens and Disease*, 2021; 18: 841–858.
15. Cason, E.E. et al. High-resolution serotyping reveals Salmonella surveillance challenges in the turkey industry. *Journal of Food Protection*, 2024; 87(10): 100319.
16. Hartono, D. et al. Impacts of particulate matter (PM2.5) on the behavior of freshwater snail *Parafossarulus striatulus*. *Scientific Reports*, 2017; 7: 644.
17. Health and Environmental Effects of Particulate Matter (PM). https://www.epa.gov/pm-pollution/health-and-environmental-effects-particulate-matter-pm
18. Zhao, J., et al. Farmed fur animals harbour viruses with zoonotic spillover potential. *Nature*, 2024; 634: 228–233.
19. Haddrell, A., et al. Ambient carbon dioxide concentration correlates with SARS-CoV-2 aerostability and infection risk. *Nature Communications*, 2024; 15: 3487.
20. Fox, M.W. *Laboratory Animal Husbandry: Ethology, Welfare and Experimental Variables*. Albany, NY, State University of New York Press, 1986.

9 Veterinary Bioethics and Planetary Health

The science and practice of human and veterinary medicine are being redefined by such global issues as climate change, rising world hunger and poverty, the many complexities of zoonotic and other infectious and contagious diseases, species and habitat extinction, and food safety and security in times of terrorism and war. The linkage of such issues and concerns is bringing human and veterinary medicine together in new ways that are being referred to in the professional literature as One Health.

RESOLVING A VETERINARY DILEMMA: PEOPLE OR ANIMALS FIRST?

A major bioethical dilemma facing the veterinary profession today is how to respond to a growing public demand for improved treatment of animals in society, especially those that are exploited for commercial purposes, from biomedical research to food production, and the mass production of purebred dogs in puppy mills, as well as the mistreatment of wild animals in circuses and zoos. There would be no dilemma if economic concerns took second place to animal

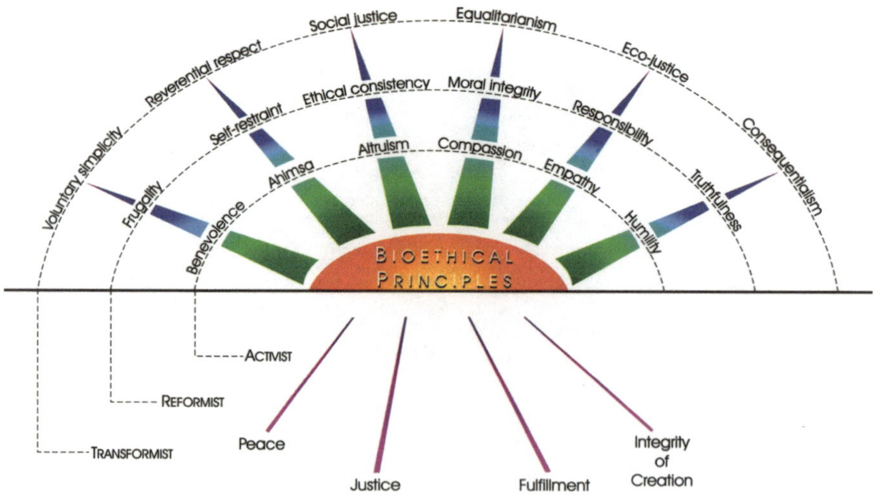

FIGURE 9.1 The scope of bioethics expands our perception and areas of planetary and social responsibility.

DOI: 10.1201/9781003569787-9

welfare and if sectors of the veterinary profession were not aligned with those vested interests in maintaining and expanding an increasingly global market economy based on animal exploitation. Fundamentally it is the consequence of anthropocentrism that leads us to regard and treat other animals as inferior. This perception is now becoming more 'zoo-centric,' where animals are appreciated and respected in their own light. This means that making interpretations of their emotional and motivational states that seem anthropomorphic are no longer taboo (1,2).

The domestication and commoditization of once-wild animals, who have not all lost their entire ethos or original natures and are therefore not yet adapted to the kinds of environments and ways of treatment to which we subject them for pecuniary ends, is a long-neglected bioethical issue. It is also a conundrum. How can we claim as a society to care for animals when economic interests take precedence over animals' welfare and overall well-being?

That our domesticated animals' ancestors were taken originally from the wild for the purpose of domestication is a fact that we must all consider when addressing animals' adaptive capacities. The welfare and well-being of domesticated animals are compromised when the highly productive varieties of pigs, cattle, poultry, fish, and other species, all genetically altered to varying degrees, are forced to live in totally unnatural, stressful conditions (3). These can range from almost total physical restriction and environmental impoverishment, as is the case for breeding sows and most veal calves (ditto many wild animals in zoo and circus cages), or extreme overcrowding, as is the case with battery caged laying hens, farmed salmon in floating net-cages, and piglets in confinement stalls and pens (4).

The narrow view of human health as being the absence of disease has been long redundant with the World Health Organization. Likewise, the narrow view that the absence of disease in animals, especially those that are raised for human consumption, is a cardinal indicator of health and overall well-being is untenable. Their health and productivity depend so much on the use of vaccines and an armament of drugs that can have serious adverse environmental, public/consumer health, and long-term economic consequences. Such treatments are not based on sound science, but on the animal productivity paradigm of agribusiness economism that is devoid of any bioethical framework. A blatant example of economic concerns trumping animal welfare concerns is in the transportation of pigs to slaughter, where the economies of transportation justify extreme overcrowding that can result in some economic losses when some pigs succumb to the stress and their carcasses (so-called 'slimy cutters') become unfit for human consumption. Likewise, the economies of scale justify large dairy herds, huge hog and poultry factories, and massive beef and dairy feedlots. But the costs in animal welfare and health, as well as the environmental and increasing public health costs, are high (5,6).

To reason that antibiotic and other drug and vaccination-maintained confinement sheds containing thousands of pigs or poultry are acceptable from an animal welfare perspective because the animals are productive (the pigs and poultry grow quickly, and the hens lay many eggs) and disease incidence is low is patently absurd. Animal health determination includes psychological as well as physical well-being, not simply the absence of disease. Animal health and animal welfare are codependent. One cannot be taken away without affecting the other. Likewise, animal welfare standards must consider not only animals' physical needs and requirements but also the related emotional, behavioral, social, and environmental needs (7).

The concept of One Health that is emerging as veterinary and human medical fields converge and collaborate with particular emphasis on environmental – indeed planetary – health, will only succeed if equal consideration is given to the treatment and alleviation of the symptoms of disease as it is to the prevention of harm and suffering to all sentient beings that contribute to the functional integrity of the Earth's ecology, as well as to the life and beauty of the Earth. We have been slow to learn that when we harm the bacteria in our digestive systems and in the soil, we harm ourselves and our crops.

Disease (and stress) prevention go way beyond better vaccinations and diagnostics, to examining our values and relationships, as in how the land and farmed animals are treated, along with the crops and foods we and they consume. Preventive medicine must also examine, like the shamans and healers of past civilizations, the icons and totems associated with the more destructive and harmful dimensions of the cultural ethos and human psyche. Physician Albert Schweitzer's remedy, which he prescribed for many of the world's ills, was based on such a holistic view of human well-being – namely, reverence for all life (8).

GLOBAL ISSUES AND BIOETHICAL PERSPECTIVES

Part of the now global, systemic pathology of increasingly dysfunctional terrestrial and aquatic ecosystems parallels the condition of local and national resources and economies (9,10). The catalysts and vectors of this pathology are multiple and synergistic (11). These include consumerism, industrialism, and human and domestic animal population expansion and concentration that combine to accelerate climate change, the loss of biodiversity, and the undermining of water and food quantity, quality, safety, and security (12).

The new 'life science' of genetic engineering biotechnology with its patented genetically modified (GM) seeds and transgenic and cloned livestock, seeking to capitalize on the world hunger and climate change crises, is more likely to worsen the situation. This is not only because GM crops are genetically unstable and so can spontaneously mutate to produce potentially toxic new protein compounds

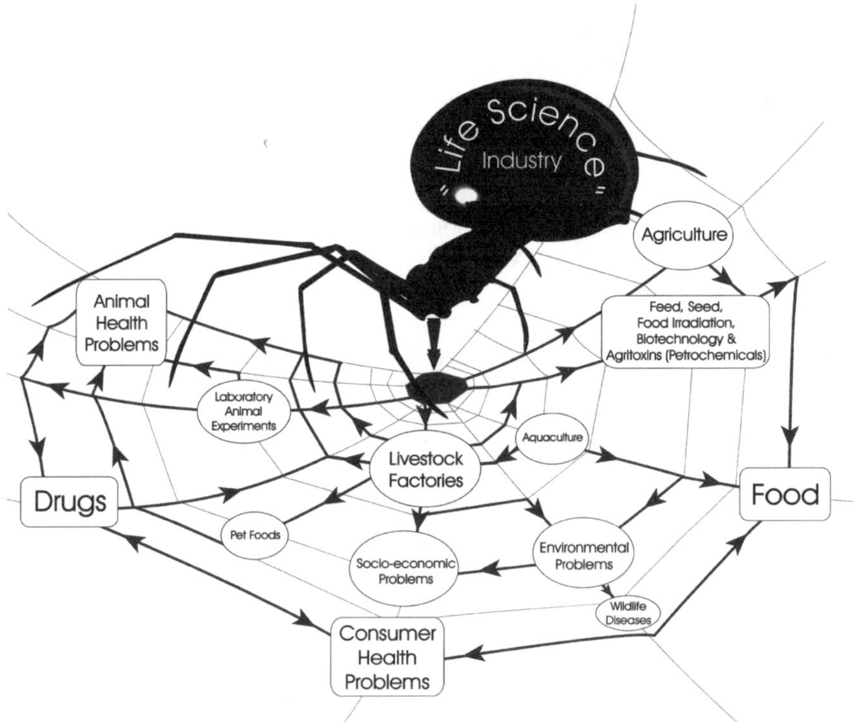

FIGURE 9.2 A conceptual visualization of the contribution of the life sciences to the exploitation of animals for food and drug manufacturing and the detrimental health consequences of this industrial complex.

(13,14). The application of this biotechnology is not consonant with the ethics and practices of more sustainable and humane farming systems (15), many of which, in contrast to conventional agricultural practices, are economically more viable and produce more nutritious food with less harmful and generally beneficial environmental consequences (16).

The contributions by some segments of the veterinary profession have a disastrous legacy. By direct and indirect association globally with the livestock industry and commodity crop producers, they are in part responsible for climate change, loss of biodiversity, and zoonotic diseases like swine and avian flu, *Escherichia coli*, and *Salmonella*. Their combined roles in contaminating the food chain, oceans, and rainwater with petrochemical fertilizers, pesticides, genetically modified organisms (GMOs) and veterinary pharmaceuticals are a matter of biological record – a legacy that will endure for generations (17). Many of these contaminants and pollutants are endocrine disrupting, carcinogenic, teratogenic, mutagenic, neuro- and cytotoxic immunosuppressants.

Not all modern medical science and veterinary practices are ill informed. Both conventional and complementary, holistic, alternative, and traditional (indigenous/native) agricultural, veterinary, and comparative medical disciplines and traditions are being utilized as valuable resources of knowledge and application to address some of the concerns and issues. Other scientific disciplines and practices are being resourced also, such as environmental toxicology, immunology, and reproductive and molecular biology. Nutritional genomics for companion and farmed animals and applied ethology/animal behavior, essential to ensure optimal well-being (health and welfare) for all captive wild and confined domestic animals, especially those used in food production, biomedical research, and for other commercial purposes, are being integrated into veterinary practice and undergraduate veterinary education. As Einstein cautioned, 'The problems that exist in the world today cannot be solved by the level of thinking that created them' (18).

EVOLUTIONARY ADAPTATION OR EXTINCTION

From a holistic perspective, it would be accurate to say that we and the world are one, being embedded in a co-creative matrix of mutually enhancing symbioses (19), from the bacteria in our soils and digestive and dermal systems, to the sustainable economies of natural ecosystems and organic and biodynamic farming practices. Some scientists name this evolutionary period the Anthropocene epoch, since the time we discovered how to make fire, tools, and weapons to alter and exploit the world's habitats. This epoch is chronicled by the ages of pyrotechnology, mechanical, chemical, atomic, and genetic engineering technologies and industries. The global crises that we face today can be seen optimistically as evolutionary challenges to our biology and psychology, shaping human nature for better or for worse and for generations to come. When we harm the Earth, we harm ourselves, and when we demean other sentient beings, we demean our own humanity.

The lack of any unified sensibility in our regard for and treatment of animals and nature – the natural creation – means, in terms of child development, a schizophrenogenic environment for character formation and personality development. (For a more detailed discussion see Chapter 12.) This may result in ethically inconsistent, morally compromised or inverted, and emotionally conflicted perceptions and treatment of other sentient beings in adulthood. Totemic and iconic values and perceptions of animals and nature range from the instrumental to the sentimental; from being objects of property, exploitation, and commerce to subjects of affection, concern, and communion; and from treating others as ends in themselves to using them as a means to one's own exclusive ends. Animals once feared or revered are now regarded variously as commodities, test subjects, objects of scientific curiosity, and potential economic or medical utility; as an indicator and flagship species of ecological well-being; and as beloved companions and family members.

The 'golden mean' of mutually enhancing symbiotic and commensal relationships is the iconic template for humane and responsible planetary trusteeship, and we have far to go ethically and legally before this state of being, where love and duty are one, is achieved. Such an achievement would herald the coming of what Fr. Thomas Berry calls the Ecozoic age (20), or, in the sense of this chapter, the Ethicozoic age, where bioethics temper the anthropocentrism of the earlier industrial age of this Anthropozoic epoch.

Reverence for all life, in practice, means equalitarianism – giving all living beings equally fair consideration – especially in terms of the consequences of human values, appetites, actions, demands, and aspirations. It is synonymous with eco-justice. The new animal welfare legislation in the UK that mandates the 'duty of care' to animals under our dominion is a positive sign of the dawning of the Ethicozoic age.

Physician V.R. Potter (21) first coined the term 'bioethics' to link the biological sciences with ethics to demonstrate how bioethics can serve as a bridge linking ecology, environmental issues, medicine, and public health. He was concerned that medical ethics was too narrowly focused. The same may be said of veterinary ethics when there is no linkage with ecological and environmental science and consideration in the practice and teaching of veterinary medicine. My friend, the late philosopher and distinguished professor at Colorado State University, Bernard Rollin, was one of the first to bring ethics into the veterinary and animal science teaching curriculums and discussed the ethics of animal care and welfare with ranchers and other animal keepers (22). He was a pioneer in setting the ethical foundation for animal welfare that led me from animal rights to a broader conception of applied bioethics and One Health. Integrating the veterinary teaching curriculum with an empathy-based bioethical template (23,24) would do much to meet the challenges and issues related to the environment, animal health and welfare, and to the ultimate advancement of civilization.

Veterinary bioethics calls on every veterinarian to apply the bioethical principle of compassionate care in their treatment of animal patients and in the advice given to client-owners and caregivers. This helps override the situational ethics of treating animals kept as commodities on factory farms where optimal care of animals on an individual basis is not normally provided for reasons of cost and where a companion animal is not given optimal care because the owner has limited financial means or does not feel that the animal is worth the expense of costly diagnostic and treatment procedures.

Rather than compromising their professional standards and integrity in such situations, veterinarians have a moral obligation to advocate compassionate care regardless of the context and situational ethics in which their services are required. This is because the bioethics of compassionate care is a fundamental human responsibility and every animal's basic right. Furthermore, compassionate

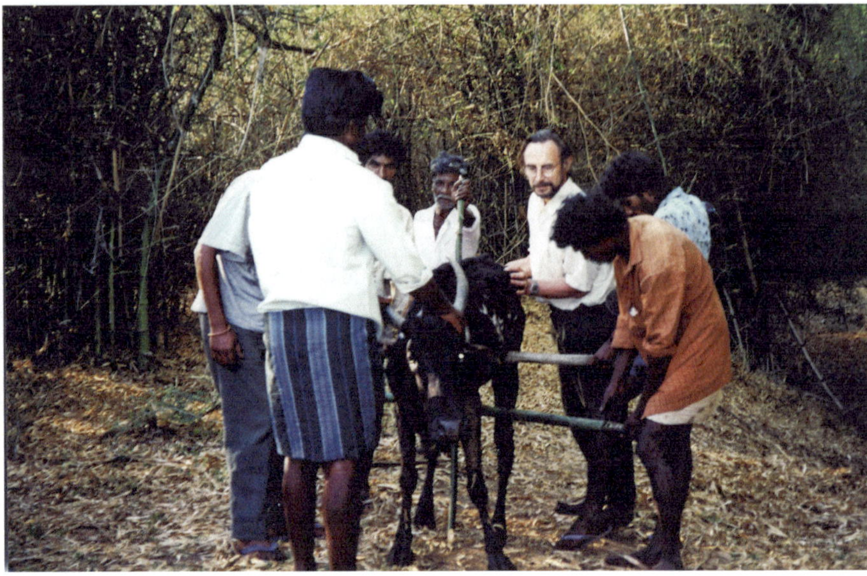

FIGURE 9.3 The author rescuing a free-ranging "scrub" cow in India, one of thousands competing with wildlife for feed, spreading disease and being killed by predators that cattle owners then kill with poisoned bait.

care is vital to animals' health, welfare, and physical and psychological well-being. It is therefore as essential a component of holistic and preventive veterinary medicine as caring for the land is a vital aspect of sustainable agriculture.

Other professions and business enterprises are similarly being called to accountability and responsibility, just as all of us in our personal lives must find ways to cause less harm to the natural world and to animals domesticated and wild in the process of satisfying our basic needs. To realize the long-term benefits of applying bioethics in our decision-making and consumer choices to our own health, to the economy, and to the entire life community of this living Earth means living mindfully and by the guiding principle of compassionate care.

No new laws, government oversight, or international conventions can equal the profound benefits that will come from the incorporation of the bioethics of compassionate care into every level of society – from business and education to health care and care of the elderly and the poor – as well as in our everyday lives, and in all our relationships, most especially with animals who continue to serve our needs in countless ways and contribute more to the greater good than all of us combined. More discussion of the challenges and future of the veterinary profession in a changing world is warranted and will be provided in the next chapter.

REFERENCES

1. Balcome, J. *Pleasurable Kingdom: Animals and the Nature of Feeling Good.* New York, Macmillan, 2006.
2. Bekoff, M. *The Emotional Lives of Animals.* Novato, CA, New World Library, 2007.
3. Webster, J. *Limping Towards Eden.* London, Blackwell, 2005.
4. Fox, M.W. *Eating with Conscience: The Bioethics of Food.* Oregon, New Sage Press, 1997.
5. Hu, F.B. and Willett, W.C. *The Relationship between Consumption of Animal Products (Beef, Pork, Poultry, Eggs, Fish and Dairy Products) and Risk of Chronic Disease: A Critical Review.* Report for the World Bank. Cambridge, MA, Harvard School of Public Health, 1998.
6. Campbell, T.C. *The China Study: The Most Comprehensive Study of Nutrition Conducted, and the Startling Implication for Diet, Weight-loss and Long-term Health.* Dallas, TX, Bell Bella Books, 2005.
7. McMillan, F.D. (ed). *Mental Health and Well-Being in Animals.* Ames, IA, Blackwell, 2005.
8. Schweitzer, A. *The Teaching of Reverence for Life.* New York, Holt, Rinhart and Winston, 1965.
9. Goodland, R. Environmental sustainability in agriculture: Diet matters. *Ecological Economics*, 1997; 23: 189–200.
10. Imhoff, D. and Baumgartner, J.A. (eds). *Farming and the Fate of Wildlife.* Healdsburg, CA, Watershed Media, 2006.
11. Diamond, J. *Collapse: How Societies Chose to Fail or Succeed.* New York, Penguin Books, 2005.
12. Korten, D. *The Tyranny of the Global Economy.* West Hartford, CT, 1995.
13. Wilson, A.K. et al. Transformation-induced mutations in transgenic plants: Analysis and biosafety implications. *Biotechnology and Genetic Engineering Reviews*, 2006; 23: 209–226.
14. Domingo, J.L. Toxicity studies of genetically modified plants: A review of published literature. *Critical Reviews in Food Science and Nutrition*, 2007; 47: 721–733.
15. Fox, M.W. *Killer Foods: What Scientists Do to Make Better Is Not Always Best.* Guilford, CT, Lyons Press, 2004.
16. Cooper, J. et al. (eds). *Handbook of Organic Food Safety and Quality.* Cambridge, UK, Woodhead Publ. Inc., 2007.
17. Steinfeld, H.P. et. al. *Livestock's Long Shadow: Environmental Issues and Options.* Rome, Food and Agriculture Organization, 2006.
18. Einstein, A. *Ideas and Opinions.* Carl Seelig (ed), New York, Crown Publishing, 1954.
19. Margulis, L. *The Symbiotic Planet: A New Look at Evolution.* New York, Basic Books, 1998.
20. Berry, T. *The Dream of the Earth.* San Francisco, CA, Sierra Books, 1988.
21. Potter, V.R. *Bioethics: Bridge to the Future.* Englewood Cliffs, NJ, Prentice Hall, 1971.
22. Rollin, B.E. *An Introduction to Veterinary Medical Ethics: Theory and Cases.* Ames, IA, State University Press, 2006.
23. Fox, M.W. Veterinary bioethics: Eco-veterinary and ethnoveterinary perspectives. *Veterinary Research Communications*, 1995; 19: 9–15.
24. Fox, M.W. Principles of veterinary bioethics. *Journal of the American Veterinary Medical Association*, 2006; 229: 666–667.

10 The Challenges and Future of the Veterinary Profession

More women than men now graduate from veterinary colleges than when I graduated in 1962. My education in England (after winning a county scholarship) was free. Today, graduates have student loan debts of around $350,000 – not a secure way to start a professional life dedicated to providing the best and most cost-effective care for people's beloved cats, dogs, and other animals they keep, often inappropriately, as 'pets.' The clinical challenge to help their animal patients are many and considerable, along with rising costs of medications and essential diagnostics. Most of their animal patients are neutered, variously confined, and often living long hours alone, fed highly processed manufactured foods and given over-the-counter antiflea and other antiparasitic drugs on a routine basis. Many dogs and ever more cats also have genetic abnormalities because of selective breeding and in-breeding that make them more prone to various diseases, vaccinations, and anesthesia, which could lead to frivolous malpractice suits and rising liability insurance costs. We can add having to address the demands and consequences of declawing cats, ear cropping and tail docking of dogs, lack of owner/caregiver understanding of animals' basic needs, and PTSD in abused and traumatized shelter animals. Highly trained and dedicated animal doctors are seeing a decline in patient treatment hours as the foundering economy affects more and more people, the core middle class and secure working class of a once relatively sustainable and equitable economic and political system.

Pet health insurance (for those who can afford it) and franchised veterinary services are now a reality. But in my estimation, the future would be more secure and the greater good better served by the professional veterinary sector that provides companion animal veterinary services to be more deeply involved in local animal shelter, animal rescue, cruelty, and protection concerns. Those in the veterinary sector would also do well to promote preventive health care, animal behavioral counseling, and even speak to schoolchildren about animals and their proper, responsible care, humane ethics, animals' interests, and rights. The many dogs who bark for hours in my and other neighborhoods and the free-roaming cats killing songbirds and chipmunks in front of us while we sit out on our decks are symptomatic of some major animal health, welfare, and care issues that have not been fully addressed and rectified. In part this is due to a lack of veterinary activity in community relations and the need for more effective municipal animal care and protection regulations.

DOI: 10.1201/9781003569787-10

A more socialized rather than overcapitalized companion animal veterinary medicine is called for, and many veterinarians are responding to that call. There is no group of professionals more qualified to share their experiences through volunteering and offering their expertise by beginning an ongoing educational dialog with animal lovers in their communities More veterinary students are also being exposed to this community-based orientation by helping in animal shelters and engaging in spay–neuter, vaccination, behavioral counseling, and adoption programs and caring for the companion animals of the poor and the homeless.

Others elect to invest in postgraduate specialist training and work in referral practices for those clients who can afford to see a veterinary dermatologist, neurologist, or other specialist for their animal's particular health issue. But many health issues may soon be greatly reduced thanks to advances in veterinary preventive and holistic, integrative medicine. Public education of the value of microchipping cats and dogs and of adopting an animal rather than purchasing a purebred or 'designer-bred' puppy or kitten also contributes to improvement in companion animal health and welfare. Many people now take their pups to puppy classes to socialize their animals and to learn animal communication and control, while cat owners learn how and why to encourage their cats to enjoy indoor life and never allow them to roam free outdoors.

FIGURE 10.1 Cats are happy staying indoors when given an enriched environment including scratching posts, toys, and playful human interaction.

FIGURE 10.2 The poorly regulated pet food industry has a wide variety of prepared foods for companion animals that are only too often unhealthful, creating a secondary market for profitable 'prescription' diets.

The old saying, 'the best medicine is prevention,' still rings true. It calls for annual wellness examinations for all animal companions and a more holistic, integrative approach to companion animal health in this 21st century. This philosophy has already led to a reevaluation of vaccination protocols, of feeding highly processed 'junk' commercial pet foods, the 'necessity' of spaying and neutering, and of overmedicating, especially with so-called preventive medications like those sold in pet and drug stores to kill internal parasites and fleas and ticks.

EXAMINATION OF POTENTIAL CONFLICTS OF INTEREST

The relationships between the corporate sector, especially drug companies, private medical practitioners, hospitals, and medical schools, are being called into question by the Institute of Medicine in the United States (1). Is a similar examination called for in the veterinary sector, where comparable corporate interests may be at play, which could affect the quality of care and services animal patients receive? There has been a lack of due diligence over the role of diet, specifically highly processed pet foods (2), in many contemporary health problems of companion animals. The same may be said about the routine application of so-called 'preventive' treatments such as antiflea and antitick topical products that only now are being fully evaluated by the Environmental Protection Agency (3). Was due diligence also lacking, in part because of inadequate information and understanding, with dog and cat vaccinations? Until recently, veterinarians strongly encouraged the universal protocol of giving dogs and cats annual 'booster' injections of multivalent live and genetically engineered vaccines.

Corporate sector partnering in academia even includes chairs and professorships named after the donating company at many veterinary colleges. What role such partnering may play in contributing to the grave consequences of poor diets, over-medication, and hyper-vaccination in companion animals by deferring to vested interests and by claiming lack of scientific proof of harm from such practices is an open question. Academia should not be exploited to garner public credibility, nor should the marketplace become the final arbiter of what is acceptable.

Examining possible conflicts of interest may be difficult, considering the partnership of the American Veterinary Medical Association (AVMA) with Fort Dodge and Merial pharmaceutical companies and Hill's Pet Nutrition, who together pledged $4.5 million in support of AVMA programs and services over a 4-year period (4). But this difficulty could become a confluence of interests once the health and well-being of companion animals are first and foremost on the agenda. The content of both the JAVMA and its equivalent with the British Veterinary Association's (BVA) *Veterinary Record* increasingly addresses issues concerning animal health and welfare, including nutrition and vaccinations.

In the UK, the government and the BVA have chosen to focus on the health and welfare problems of a genetic origin, primarily in pedigree dogs. Some critics believe that this is a massive displacement, since it is the genetic susceptibilities to dietary diseases and vaccinoses (adverse vaccination reactions) in specific breeds that make them the canaries for the canine population at large (5). The appropriate use of vaccines (6), various 'preventive' veterinary drugs, prescription diets,

FIGURE 10.3 The breeding of dogs with extreme abnormal body shapes and flat faces (brachycephaly) and ear cropping are welfare issues the veterinary profession in many countries is now addressing.

and the adequacies of manufactured cat and dog foods also need to be considered if the mandate of the British government is to protect the health and welfare of companion and other animals and not just better regulate breeding practices. British dog breeders feel they are the scapegoats and are taking all the blame for myriad and costly health problems in today's canine population. The same can be said for the major ailments in the feline population, where poor diets and adverse drug and vaccine reactions similarly take their toll, according to Hill's former director of technical affairs, veterinarian Dr. Elizabeth Hodgkins Esq. (7).

Of course, there are confluences of interest that can benefit all, and this would be forthcoming, I believe, when there is a more integrated approach to animal health and welfare. This could be developed from a bioethical basis (7) by veterinary teachers, researchers, and practitioners and establishing a Council for Veterinary Bioethics to respond to the call for an examination of possible conflicts of interest within the profession.

One reason for the veterinary profession's current hard times and unclear future may be its historical double vision or conflicted mission. The profession seems torn between serving the best interests of animals and serving the interests and demands of its animal-owning clients and the larger society – both of which may not be in accord with those of the animals. The AVMA was founded in 1863, but its *Veterinarian's Oath* included no reference to animal welfare 'as a priority of the profession' until 2011. I have long seen the profession torn between two often-conflicting, but I believe resolvable, forces and sets of values – concern and responsibility for animal suffering, health, and welfare (the 'calling' of most veterinary school applicants) and economic realities such as the cost of services performed and the monetary, rather than the emotional (and other) value of animals. However, I do find grounds for optimism in the emerging One Health movement (8), where human health and well-being are at last being linked to animal health and welfare, as well as to environmental and ecological health. But for the One Health concept to be effectively implemented, the bioethics of equal consideration must be applied not only to humans but also to our treatment of animals and the environment. This is in accord with the moral imperatives of the animal rights and 'deep' ecology movements, long opposed by the biomedical, agricultural, and other industries and business establishments.

OBLIGATION TO ADVOCATE COMPASSIONATE CARE

From my bioethical perspective, holistic human and veterinary medicine addresses the diagnosis, treatment, and prevention of disease/illness from an integrative mind–body and environment paradigm. The triune of psyche, soma, and spirit – mind, body, and the patient's will to live – is especially important for the stressed, distressed, and chronically ill and those who are impaired physically and emotionally. Quality of life and quality of environment are paramount considerations. Far too many animals kept as 'pets' lead socially and environmentally impoverished lives in crates, cages, even chains, as do elephants and other wild species kept for public entertainment. The ways in which most animals are raised for human consumption are a horrendous reflection of human insensitivity and instrumental rationalism leading to the destruction, depletion, and desecration of ecosystems integral to the health of the biosphere and all who dwell within.

In the *Codex Veterinarius* drafted by the German Veterinary Association for the Protection of Animals (9) we find this statement:

> Veterinary action for the welfare and protection of animals is guided by the principle of respect for life and the awareness that the animal has a dignity which is to be respected. Therefore, protection and care for an animal cannot be dependent on its economic value. In situations of diverting interests between moral obligations and economical pressure it is essential to consider carefully the respect for life against the productive use of life in all its forms of existence.... Considering carefully the opposing interests and needs, the interests of the human being should not automatically be considered to be more important than those of the animal.... When in doubt veterinary surgeons should let themselves be guided by the principle: *In dubio pro animale*!

This idiomatic Latin phrase can be roughly translated as, 'When in doubt, support the animals!'

There is growing consensus that veterinarians have an ethical obligation to advocate for compassionate animal care regardless of the context in which their services are applied. This is because of an emerging recognition that compassionate care is our fundamental human responsibility and every animal's deserved and basic right. Furthermore, compassionate care is vital to animals' health, welfare,

FIGURE 10.4 Veterinarians understand animal behavior to safely handle them and win their trust.

and physical and psychological well-being. It is as essential a component of contemporary veterinary medicine as caring for the land is a vital aspect of sustainable agriculture.

VETERINARIANS AS INTERLOCUTORS

Veterinarians have been seen as a unique association for millennia. As interlocutors among people, animals, and their environments, veterinarians' roles and knowledge – both empathic and scientific – have been highly valued by societies. Indeed, according to Professor Calvin Schwabe in his book *Cattle, Priests and Progress in Medicine*, the earliest veterinarians were priest-healer members of ancient Egypt's many dynasties (10).

Today, gender-linked differences in moral philosophy may be factors associated with current shortages of veterinarians in the food animal and laboratory animal medicine sectors (11). Veterinarians in companion animal practice and those dealing with 'exotic' animals face ethical dilemmas daily. There is an urgent need for ethics educational tools and approaches specifically designed for veterinary students. The influence of vested corporate interests in veterinary schools has also increased significantly in recent years and is now being questioned as a broader ethical issue.

STUDENT DEBT BURDEN AND INCOME-CHALLENGED COMMUNITY ACCESS TO CARE

Lobbying for better funding from government, industry, and non-profit sources for student scholarships or debt forgiveness for pledging to work in much-needed sectors, such as overseas aid and development, emergency services, and in low-income neighborhoods (like the People's Dispensary for Sick Animals in the UK), may be considered enlightened professional self-interest. However, we know that at the community level within the United States, as revealed in a recent report by the Humane Society of the United States ('Pets for Life – A New Community Understanding'), there is great need to remove cost barriers to animal care and veterinary services for economically blighted communities in order to improve their community's animal health and reduce shelter overpopulation. *News From the Hill* (February 1, 2024) reported that the ongoing shortage of livestock and public health veterinarians threatens the health and well-being of farm animals as well as human health and the nation's food supply. Educational debt discourages veterinarians from practicing in rural areas and public health fields, where pay is comparatively low, and congress must pass the bipartisan Rural Veterinary Workforce Act to help ease the shortage.

This initiative is an ethical imperative so long as we continue to raise animals for food. Large factory farms in the United States, UK, and other countries put

FIGURE 10.5 There is an urgent need for veterinary services in poor rural communities around the world, to benefit the people and their animals, and to protect contiguous wildlife from diseases spread by livestock and poultry.

small producers out of business (with the 'economies of scale') and in the process impoverish rural community diversity and culture, including support and need for rural veterinarians.

TOWARD A NEW PROFESSIONAL PARADIGM

Redefining the role of the veterinary profession in these times is critical to our profession's future and to civilization itself. Putting animals and the environment first may be in the enlightened self-interest of civil society, but doing so would be an admittedly improbable paradigm shift, considering the nature of human psychology. Establishing a higher priority of concern for wild and domestic animals and the natural environment over trivial and harmful human needs and demands would be a significant step toward attaining a more sustainable economy and creating a better world for us all. I believe it is the veterinary profession's current challenge – and our duty – to apply our expertise in more effectively advocating for this new paradigm.

Anthropologist Prof. M. Guenther describes such resonance in the African Bushmen, which I believe is an innate or inborn ability of our species. Considering that for approximately 95–98% of our time on Earth as humans we were gatherer-hunters, this ability could be restored from our latent cellular memory. Sometimes while treating sick animals and in making a diagnosis, I was so attuned to the animal that I felt pain or discomfort in parts of my own body that corresponded to the animal's illness or injury.

Prof. Guenther writes:

> Throughout the hunt the hunter would monitor his every thought, emotion and action, in order to sustain the bond of connectedness with the animal by which he felt he could steer the hunt towards an auspicious conclusion…. The bond of sympathy was something set up in the hours or days preceding the hunt, when the hunters would attune themselves spiritually to one animal species or another and, in the process, attempt to gather whatever presentiments they could about the impending hunt: the animals they might encounter, the direction they could come from, the likely dangers, the duration of the hunt. These presentiments … activated the hunter's entire body; they were felt at his ribs, his back, his calves, his face, and eyes. His body would be astir with the 'antelope sensation,' at places on his body corresponding with those of the antelope's. (12)

If we are to 'evolve,' we must regain a heightened sensitivity for fellow creatures and the living Earth, for when we harm others and the Earth, we harm ourselves. Part of our healing and wholeness therefore involves our nascent ability to empathize, and we must have the courage to suffer in the process before we can put compassion into action and truly love.

VETERINARIANS WELL-BEING

So-called moral distress for veterinarians and other professionals confronting on a regular basis and having to engage in or support actions contrary to their principles and values – personal and professional and under oath in some instances – is a potentially serious mental health issue and occupational health risk. This may be a contributing factor in the disturbing 2018 report that 24.9% of veterinarians in the United States have considered suicide. This report also noted that while 41% of veterinarians overall would recommend the profession to a friend or family member, only 24% of veterinarians aged 34 and younger would do so. Even those scoring high in well-being and mental health do not recommend the profession (13).

A cascade of stressors associated with running a private veterinary practice contributes to burnout, compassion fatigue, lack of job satisfaction, and low self-esteem, along with student debt loans, long work hours, and understaffing. They can erode professional expectations and aspirations. In a recent survey of veterinarians in the UK, 'all staff find working for a chain/corporate practice more stressful than an independent or private clinic' (14).

That both recent and older graduating female veterinarians had more mental health issues than males and that those in the food animal sector scored highest in well-being above all other segments of the veterinary profession are instructive findings (15). Animals in the food sector are easier to objectify as commodities without the emotional dynamics of companion animal patients and clients, which call for empathy and understanding – qualities applied more often by females than males in social interactions and relationships (16–18).

The greater the empathy, the greater the burden of others suffering which, if not addressed, may lead to compassion fatigue, burnout, and other mental health problems. Mental as well as physical health issues will continue to be occupational hazards for veterinarians, many of which are shared by other caring professions. Peer support and pure reason can help address moral conflicts and related distress, with the application of veterinary bioethics offering some insight and balance to empathy and emotional engagement.

Notable developments include the rising number of women in the companion animal sector, both in the United States and the UK (which has improved the health of companion animals), and fewer graduates specializing in the food animal sector, which leaves some rural communities with no veterinary services in the United States.

But surely, related to self-worth rather than financial remuneration is the individual's sense of purpose, service, and value to the common good. As new veterinary specialist disciplines are being recognized, so new opportunities for veterinarians are emerging in such areas as disaster relief, animal rescue, humanitarian, public health, and conservation organizations. Is there nothing more noble and needed than good people dedicated providing the best of care, health, and well-being for all animals, wild and domesticated, in the world community?

We are facing a fundamentally moral crisis in civilization today, which has been brought on by what the Hopis call *Koyaanisqatsi* – living out of balance with other species and the natural environment. Cellular and ecological dysbiosis, cultural dystopia, and mental dysphoria are coins of the same currency from the anthropogenic exchange. Climate change, cancer, and species extinction are symptomatic and consequential. There is clearly a crisis of spirit, with approximately 70% of my veterinary colleagues contemplating suicide in these times. Compassion fatigue and moral distress in this and other healing and caring professions and avocations will continue to take their toll until antidotes are found (19). The antidotes of enthusiasm, courage, and commitment of veterinarians to adopt a more holistic approach to animal health and disease treatment and prevention and ultimate engagement in the One Health movement are considerable. Such involvement is also an antidote of reconciliation between what one has been taught and made to believe and what is real. Sound science, evidence-based medicine, and integrative preventive health-care maintenance must be coupled with bioethics and 'planetary CPR' (conservation, protection, and restoration) for the common good, upon which our ultimate well-being and security depend. I was encouraged by this statement by two veterinarians:

> The world is losing wildlife species at an unprecedented rate. Habitat destruction, overexploitation, and pollution are the leading causes of biodiversity decline. As a threat multiplier, climate change exacerbates these processes as demonstrated by the death of several billion wild animals in the last few years from wildfires, floods, heatwaves, and other natural disasters. In the face of these challenges, veterinarians have unique and important skillsets to contribute to wildlife conservation and the preservation of biodiversity at many levels. Veterinarians can organize and train to mobilize wildlife extraction, rescue, and rehabilitation units during natural disasters as well as build relationships with rehabilitators to provide their services for general wildlife rehabilitation needs. They can work in transdisciplinary teams to provide veterinary expertise for ecosystem health and rewilding projects. They can become sustainability champions by providing pollinator and wildlife friendly habitats at their clinics and reducing clinic waste and energy consumption, and they can engage in science communication and advocacy. When provided with the necessary information, resources, and action items, veterinarians can increase their positive impact and personal well-being through purposeful, value-driven, community-building efforts to support wildlife conservation and biodiversity. (20)

The next chapter will expand on the role of veterinarians and allied professionals in leading the way to One Health.

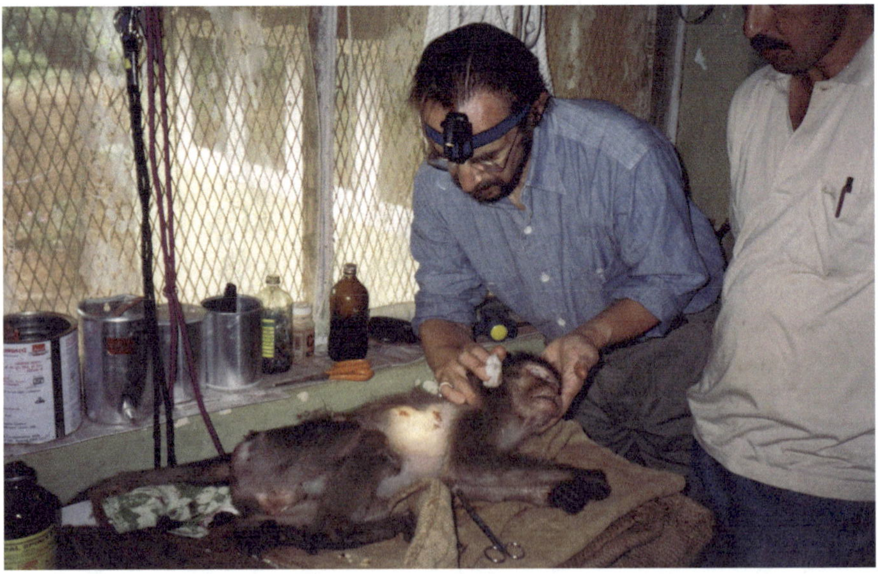

FIGURE 10.6 Veterinarians can play a significant role in treating injured and diseased wildlife around the world, such as this Indian bonnet macaque being treated after being bitten by a wild dog, or dhole.

REFERENCES

1. Sternbrook, R. Controlling conflict of interest – Proposals from the Institute of Medicine. Published at www.nejm.org, April 29, 2009. https://doi.org/10.1056/NEJMp0810200
2. Fox, M.W., Hodgkins, E. and Smart, M. *Not Fit for a Dog: The Truth about Manufactured Dog and Cat Food*. Fresno, CA, Quill Driver Books, 2009.
3. AVMA News. Topical flea and tick products come under EPA scrutiny. *Journal of the American Veterinary Medical Association*, 2009; 234: 1228.
4. AVMA News. AVMA enters into multimillion-dollar partnership with companies. *Journal of the American Veterinary Medical Association*, 2008; 233: 219.
5. Dodds, W.J. Vaccination protocols for dogs predisposed to vaccine reactions. *Journal of American Animal Hospital Association* 2001; 37(3): 211–214.
6. Schultz, R.D. et al. Titer testing and vaccination: A new look at traditional practices. *Veterinary Medicine*, 2002; 97: 1–13 (insert).
7. Hodgkins, E.M. *Your Cat: Simple New Secrets to a Longer, Stronger Life*. New York, Thomas Dunne Books, 2007.
8. Fox, M.W. Veterinary Bioethics. In Schoen, A.M. and Wynn S.G. (eds.), *Complementary and Alternative Medicine* (pp. 673–678). St Louis, MO, Mosby, 1998.
9. Burns, K. Spreading the one-health concept. *Journal of the American Veterinary Medical Association*, 2012; 240: 112–115.
10. Burgermeister, S. and Fikuart, K. Ethical guiding principles for veterinary behaviour with respect to the welfare and protection of animals. *ALTEX*, 1998; 15(4): 209–212.
11. Schwabe, C. *Cattle, Priests and Progress in Medicine*. Minnesota, University of Minnesota Press, 1978.
12. Narver, H.L. Demographics, moral orientation, and veterinary shortages in food animal and laboratory animal medicine. *Journal of the American Veterinary Medical Association*, 2007; 230: 1798–1804.
13. Guenther, M. *Tricksters and Trancers: Bushman Religion and Society*. Bloomington, IN, Indiana University, 1999.
14. Batchelor, C.E.M. and McKeegan, D.E.F. Survey of frequency and perceived stressfulness of ethical dilemmas encountered in UK veterinary practice, *Veterinary Record*, 2012; 170: 19–20.
15. Nolen, R.S. Mental health, well-being problem serious, not dire: Study. *Journal of the American Veterinary Medical Association*, 2018; 252: 392–393.
16. News and reports. Salaries and stress: How are they linked? *Veterinary Record*, 2018; 182: 65.
17. Gilligan, C. *In a Different Voice*. Cambridge, MA, Harvard University Press, 1982.
18. Christov-Moore, L. et al. Empathy: Gender effects in brain and behavior. *Neuroscience and Biobehavioral Reviews*, 2014; 46: 604–627.
19. *AAHA's Guide to Veterinary Practice Team Wellbeing*. American Animal Hospital Association, 2018.
20. Cerda, J.R and Webb, T.L. Wildlife conservation and preserving biodiversity: Impactful opportunities for veterinarians. *Journal of the American Veterinary Medical Association*, 2023; 261(7): 1077–1085.

11 Veterinary and Allied Professions
Leading the Way to One Health

The only thing worse than being blind is having sight but no vision.

–Helen Keller

We are part of Nature, our ultimate provider. If we respect and work together to protect the environment – the land, the waters, and the air we share – as well as all our relations, plant and animal, Nature will not become our ultimate nemesis…. In wildness is the preservation of the world.

–Henry David Thoreau

As long as people will shed the blood of innocent creatures there can be no peace, no liberty, no harmony between people. Slaughter and justice cannot dwell together.

–Isaac Bashevis Singer

Until he extends the circle of his compassion to all living things, man will not himself find peace.

–Albert Schweitzer

Nature, the natural world as we know it today that has been variously affected by humans in the recent and distant past, is as much our apotheosis and nemesis as our provider, teacher, healer, and inspirer. This all depends on our state of mind – how we perceive, value, relate, and treat other sentient beings, plant and animal, and their habitats that comprise our life-sustaining environment we share and continue to pollute and defile. American social critic Marya Mannes wrote, 'The earth we abuse and the living things we kill will, in the end, take their revenge; for in exploiting their presence we are diminishing our future.' British statistician and economist E.F. Schumacher stated that 'Modern man does not experience himself as a part of Nature but as an outside force destined to dominate and conquer it.' We have surely reached the end of the beginning of this anthropocentric, Anthropocene era and must now choose to either evolve or continue along the path of suffering, destruction, and extinction of the life and beauty of our planet if we do not achieve a more eco-centric, empathic perception and way of life.

DOI: 10.1201/9781003569787-11

FIGURE 11.1 Monsanto/Bayer's Roundup (glyphosate) herbicide is harmful to soil microorganisms, depletes crops of some nutrients, harms wildlife, and is a significant health hazard to farm applicators and consumers.

Once all rivers ran clear, free, and clean and the pure, sweet air sang with the insects and birds through forests and grasslands. The environment provided food for the human spirit, body, and soul, as was all else the good Earth provided, engendering in our ancestors an abiding sense of thanksgiving, kinship, and reverence for both plant and animal beings. That was long before the enslavement of man and beast; of genocide, ecocide, the slave and fur trades, the timber, oil and cattle barons, moguls of mining and industry, and the relentless drive of conquest, colonialism and empire building under political, economic, religious, and military hegemony.

Once the springs from mountain valleys were pure, but now the water is contaminated from chemicals in the rain, air, and snow that the forested slopes and peaks collect. These chemicals are no longer neutralized by the microorganisms and mineral filters in the valley floor, because most have been destroyed and disrupted by human agricultural, mining, and other activities performed in ignorance of how nature can help us. Our lack of respect for the basic elements of life – water, soil, and air – has adverse consequences worldwide. Declining soil productivity and air and water quality affect human health and the global economy. Climate change has brought unquenchable forest fires from continent to continent as oceans rise and, in warming, provide more energy for tropical storms now increasing in frequency, intensity, and duration. According to the United Nations report *Groundswell*, 200 million people may be climate change migrants seeking refuge from homelands made uninhabitable by floods and droughts by 2050.

DYSTOPIAS, DYSBIOSIS, AND EXTINCTIONS

If industrial civilization had embraced the Great Law of the Iroquois Federation (the Haudenosaunee), which advocates thinking seven generations ahead (about 140 years) to determine whether the decisions made today would benefit their children living that far into the future, it would probably not be facing its own extinction today. The Earth has undergone five major extinctions in the last 500 million years. Except for the Cretaceous extinction event, these were primarily caused by dramatic changes in the concentration of atmospheric greenhouse gases (CO_2 in the case of past extinctions). Too little, and the Earth freezes; too much, and it overheats, acidifying the oceans in the process. A dramatic reduction in species accompanied all five, but the Permian event was the worst, with 90% of life lost and the Earth perilously close to sterilization.

We are currently experiencing a sixth major extinction. Like the others, the cause is an imbalance of CO_2 and other greenhouse gases. Unlike the others, the trigger is a single species, humans. It is why many scientists have unofficially dubbed the current epoch the Anthropocene. Past extinctions occurred over tens of thousands to millions of years, an eternity by human standards, enabling many species time to adapt and evolve. The current extinction is happening over the time scale of decades to hundreds of years. Estimates suggest that we are injecting greenhouse gases into the atmosphere at ten times the rate they were being added naturally during the Permian extinction. Any curtailing of this rate at this point, although admirable, is likely to be insufficient to prevent catastrophic and lasting climate events. The damage that is already done (e.g., to the coral reefs) will certainly not be repaired naturally for many human generations. Another critical concern calling for immediate regulation and oversight is the escalating energy and fresh water use to maintain data storage centers that are proliferating to provide cryptocurrency and artificial intelligence (AI) services, which many contend will exacerbate climate change.

OUR ANIMAL RELATIONSHIPS

The Anthropocene extinction refers to the demise of many species and loss of biodiversity caused by our impact on the environment. Reptiles and amphibians are cold-blooded (poikilotherms), which limits their ability to cope with sudden changes in temperature and humidity. We and other warm-blooded, homeothermic animals can cope better. These stressed poikilotherms are now being decimated by fungal infections, indicative of impairment of the reptiles' immune systems to ward off infection.

Taking steps to prevent further biodiversity loss and stem this Anthropocene extinction of species calls for international government collaboration and science-based responsible planetary stewardship. When we lose species, ecosystems become dysfunctional, allowing noxious and invasive species to flourish and pose a threat to agriculture, food security, and public health.

The significance, value, and role of both wild and domesticated animals in various cultures include spiritual (totemic and symbolic), biological and ecological (scientific), emotional (affection/aversion), and commercial (including consumptive) perspectives. Such diversity in perception and related treatment is the antithesis of a unified sensibility based on respect for all life and valuing all living beings equally and fairly. This equality extends the Golden Rule to temper the harmful consequences and historical legacy of anthropocentrism's limited and self-limiting worldview.

Ethical consistency in our treatment of all beings based on the wisdom of the Golden Rule (avoiding causing harm to others that we would not wish for ourselves) is enlightened self-interest; when we harm others, we ultimately harm ourselves. For example, the sciences of ecology and molecular biology have revealed the vital importance of bacteria and other microorganisms in the living soil and in our digestive systems, the former being essential for plant growth and health and the latter being intimately connected to our immune, endocrine, neurological, cognitive, and other systems and functions. The complexities and interdependencies of living systems and life processes are challenges to the sciences and to human understanding. It is also important to holistic, integrative medicine as well as to farming practices (e.g., aquaculture and agroforestry) and other industries and human activities that affect these systems and processes. Proceeding with the precautionary principle as our compass to direct us away from hubris, rationalized selfishness, and denial, human progress is a possibility and not a false hope.

Contemporary society has intensified the conflict of interest for the veterinary profession by first endorsing as accepted norms various extremes in selective animal breeding (with genetic anomalies and compromised phenotypes); in keeping wild animals as 'pets' and exploiting them for circus and other commercial purposes; and in animal husbandry practices, especially in factory farms/concentrated animal feeding operations, which are contrary to the best interests of the animals and to their health and well-being. These are the primary concerns and responsibilities of the veterinary profession, yet the various animal industries and animal-owning public essentially dismiss these concerns of veterinary bioethics (1) as contrary to the accepted norms of animal use. The underlying economic and cultural justifications of what often amounts to abusive animal exploitation have put the veterinary profession in a double bind. By not asserting the core principles of veterinary bioethics and animal welfare science and ethics and being a guiding influence in society, the profession has become increasingly marginalized by the economics of the livestock industry. While it strives to serve this and the pet, exotic species, and other animal industries, it is seeing revenues decline as the industrial economy seeks more 'efficiencies' and continues to flounder.

FIGURE 11.2 Alternatives to the use of animals in product safety testing and drug and vaccine manufacture are long overdue.

It is encouraging that informed individuals outside of conventional animal-related circles such as Dan Imhoff, musician and artisan food producer, has brought wider public attention to animal farming practices in his book, *Farming with the Wild: Enhancing Biodiversity on Farms and Ranches*, and in *CAFO: The Tragedy of Industrial Animal Factories*, which he edited, and *Farming and the Fate of Wild Nature: Essays on Conservation Agriculture*, which he co-edited with Ann Baumgartner.

Millions of rural families in India and other developing countries, including some who are forced to move into urban slums, are in part sustained by an often-shared milk cow and by a few goats and chickens. Few are fortunate to have access to grazing land, with much communal land being overgrazed as per ecologist Garret Hardin's (2) predicted global 'tragedy of the commons.' Without urgently needed subsidized veterinary services and supplies of animal feed and water (rather than scavenging off garbage), this sector's farmed animal population (including bovids and equids used for draft work) continues to be a neglected and significant incubator of diseases harmful to other farmed animals, to wildlife, and to the human populace locally, nationally, and internationally. These problems are being compounded by the proliferation of intensive methods of livestock and poultry production using imported breeds. These operations become epicenters of pollution and disease and, along with imported varieties of commercial crops, accelerate the loss of biological diversity as indigenous breeds are marginalized from the agricultural system.

The veterinary profession, along with epidemiologists and ecologists, can be significant contributors of expertise to the One Health approach to the rising environmental and public health issues of food-borne illnesses and zoonotic diseases, including those arising from disrupted and endangered biotic communities and wildlife species. These professions can also help establish laws and public policies that better serve the common good and the good of the commons and all who dwell therein, on land and in the waters.

When we care for the land, the land will care for us, and when we care for animals wild and domesticated, we care for ourselves because our health and well-being reflect theirs. This is affirmed by the science and ethics of One Health (3). Three-quarters of novel or emerging infectious diseases in humans are zoonotic, and the international wildlife trade – estimated to be a $10 billion market each year – is a major source of risk. A series of public health crises have put the issue in the spotlight, shifting attention from diseases of economically important livestock species to those that are most likely to cause problems in humans. A One Health approach that accounts for the entire human–animal–ecosystem interface is the solution to mitigating risk. The consumption of exotic wild animals, wildlife farming and trade, livestock management practices, keeping exotic animals as pets, and human encroachment on wildlife habitats all put humans at risk from zoonotic diseases.

A substantial percentage of the annual profits of the pharmaceutical industry should be provided to support public health and preventive medicine initiatives and independent research in alternative, complementary, and integrative veterinary and human medicine with special emphasis on nutrition and epigenetics. The dominance of conventional/allopathic practice relying on synthetic pharmaceuticals and a new generation of costly genetically engineered biologics to treat various human and non-human medical and psychological/behavioral conditions is facilitated by alignment with the health insurance industry. The rising costs of prescription drugs and the limited ability of government to effectively guarantee drug safety and efficacy with fast-track approval of patented pharmaceuticals and others imported from countries like India and China, with endemic problems of corruption and price fixing are a major obstacle to real progress in One Health. The mechanistic, reductionistic, and economistic imperialism of conventional diagnostics and treatments in the health insurance and drug industry dominated mainstream human and animal care professions and was predicted decades ago by Ivan Illich (3).

An integral aspect of One Health is the saving not only of indigenous crop varieties and domestic animal breeds, which are generally better adapted climatically and more disease resistant than imported varieties, but also indigenous knowledge. This includes tried and true methods of animal husbandry and disease prevention and traditional medicinal treatments. One traditional medical specialty is the use of botanical remedies which, as several authors have documented (4–6), should not be dismissed as scientifically unproven and abandoned in favor of synthetic pharmaceutical products. Synthetic pharmaceuticals are often too costly and sometimes unavailable in many rural communities.

The veterinary and other professional organizations, multinational corporations, and governments alike, as well as consumers, have the power and responsibility to determine the future quality of life on planet Earth. This will be more for the better than for the worse when improving the health and well-being of both wild and domesticated animals and is recognized as a transcultural ethical imperative. This auspicious beginning is heralded by veterinarians' commitment to the One Health movement (which is more than promoting vaccinations and various pharmaceuticals), to wildlife and endangered species conservation, and to making long-overdue improvements in animal care and well-being though a holistic, integrative approach, as advocated by the American Holistic Veterinary Medical Association and other organizations and individuals of like mind.

Albert Einstein observed, 'No problem can be solved from the same level of consciousness that created it.' Only from the perspective of our being part of nature can an objective, rational, and existential assessment be made of our life's purpose and its consequences in terms of how our life choices and values affect ourselves and others for better or for worse. Solutions for most existential problems come from cooperation rather than competition, acceptance rather than denial, education rather than indoctrination, renunciation rather than conformity, passion for justice rather than retribution, and the courage to speak truth to power.

NATURAL PERSPECTIVES AND NATURE-HEALING

When we realize that we are related to each other and to all sentient beings, we discover the true meaning of freedom for all and begin to live spontaneously in accord with the Golden Rule. We all need to consider Sioux Chief Medicine Man Black Elk's statement, 'Nothing will be well unless we learn to live in harmony with the Power of the World as it lives and moves and does its work.' If there were no stones, we would have no minerals for our bones and never walk, or for our brains and never talk. If there were no trees and other green life on land and in the seas, we would not breathe without the oxygen they give. We would not be red in blood and living free without the oxygen-binding iron from Earth's magnetic core, spread by primordial volcanic activity. Without the microorganisms in healthy soils, we would have no nutritious plant foods nor our own related gut microbiome essential for our health, immune, and digestive systems. Ecology, molecular biology, and other natural sciences affirm these life-giving and sustaining relationships between the universal, humanity, and Earth.

The law of nature and nature of law are 'blind' because they are impartial and therefore objective. Only then can justice be served. The law of nature punishes us when we harm the natural systems we are connected to and depend on and when we violate natural law by overconsumption, overpopulation, overextraction, and depletion of finite resources and pollute our shared environment. Other affronts to the law of nature include harming other species by extermination, habitat destruction, and exploitation that is religiously, culturally, and economically sanctioned, which open us up to zoonotic diseases and pandemics like COVID-19. Cancer and brain and body birth defects are often a direct result of environmental contamination.

The physicians' Hippocratic maxim, *primum non nocere* ('first, do no harm') is derivative of natural law. If the big business sector were to make a similar pledge, we would surely see less social, environmental, and ecological injustice. Our affinity for and love of nature and the nature of such affinity and love are innate and intertwined. Love can link us to the sacred, to divine presence, and inspire reverence and awe.

The healing powers of nature and the nature of healing are interconnected. 'Forest bathing' and spending time in nature promote our own innate powers of psychophysical healing – what Albert Schweitzer called the 'physician within.' Efforts to protect and heal nature from human encroachment, desecration, and pollution are therefore integral to our own well-being. The biological sciences can help us appreciate and respect all life – even bacteria, viruses, and other microorganisms with whom we have co-evolved and would not and could not exist without. Some of the earliest life forms, like the cyanobacteria, generated the first oxygen on the planet by evolving photosynthesis. At least one virus played a critical role in our nervous system and brain development.

The science of nature and the nature of science are the existential reality of truth. Nature's economy is the wise human economy of regenerative sustainability. Nature's medicine is the nature of healing, which embraces more than vaccines, drugs, genetic biomedical engineering, and surgery. The ethics of nature and the nature of ethics lead us to the bioethics (as described in my book *Bringing Life to Ethics*) of sustainability: Of life giving more to life than it takes for the common good and the good of the commons, as well as the integrity of Earth's life community.

Our love of nature and nature of love are connected empathically with our innate, ancestral, and spiritual affinities and cherishing of other beings – human and non-human, animal, plant, and stone. The nature of love does not only involve loving other humans, since love cannot be exclusive or possessive without causing harm. Selfish love breaks our sacred life- and spirit-sustaining connections, as when love becomes conditional in our seeking to be loved rather than to love. There are myriad other ways that limit our lives and so often harm others and nature, mother of us all.

FIGURE 11.3 Greater public recognition of animal sentience, notably of the intelligence of dolphins, the 'mind in the waters,' is helping support international efforts to protect and conserve the last of the wild.

ANIMAL RIGHTS RESET: ONE HEALTH, ONE EARTH

Veterinarian Steven McCulloch's call for 'Vets to be strong advocates for animal rights' (7) raised an issue that has been lacking in the public voice of the profession for decades. I believe that the case for animal rights needs to be recast. It is not simply a 'moral' issue. Considering the rationalizations justifying animal exploitation and suffering by various vested interests and their lobbyists, appeal to reason and empathy may yet prevail. But only when there is also a human benefit.

It is beneficial to take steps to reduce the severity and duration of climate change. That means raising and consuming fewer animals, which would allow for rewilding and restoration and protection of wildlife and biodiversity, essential components of a healthier environment and human populace. According to the sustainable animal agriculture policy statement of the British Veterinary Association, 'Sustainable animal agriculture should be undertaken in a way that is environmentally, ethically, and economically acceptable for consumers, producers, and wider society.' From a One Health perspective, policies relating to sustainable animal agriculture must also address the use of natural resources and the protection and conservation of wild species, habitats, and biodiversity to reduce the ecological footprint of animal agriculture. The veterinary profession is beginning to promote the concept of 'less is better' in which people 'reduce their consumption of animal-derived products whilst maintaining proportional spend to buy better, higher quality goods' (8).

One major reset is called for in the widespread belief that owned cats have a right to roam free and hunt and that this behavior enhances their well-being. This can put their health at risk as well as that of songbirds, other wildlife, and public health due to several zoonotic diseases these cats could bring into their homes (9,10). The right to live free for other invasive feral domesticated animals such as pigs, goats, and cattle is not generally as sentimentalized as it is for cats. Humane methods of population control and removal or containment are scientifically/ecologically valid and socially acceptable (11).

Owners of 'exotic' pets who believe they have a right to life in the wild and set them free when they no longer wish to keep them, or accidentally allow their escape, have created problems in several countries. Such non-native animals compete with indigenous species and spread diseases. The international trade in exotic species and captive breeding must be closed immediately, except for legitimate conservation purposes. These changes will protect biodiversity and support a One Health approach in which the veterinary profession can take a leading role.

All indigenous animal and plant species and their ecosystems should have legal standing to protect them from human exploitation and other human impacts – direct and indirect – that cause suffering and loss of biodiversity. Anticruelty laws protecting some domestic species, but not all wild and domestic species, are borderline arbitrary and capricious. Existing protection laws are poorly enforced in many countries. Such protection should be provided under the laws of every country for all sentient species including 'pests,' 'vermin,' and food animals (livestock, poultry, fish and other seafood), where the most humane killing methods must be utilized when killing is justified.

Veterinarians also play a pivotal role in investigating cases of animal cruelty and neglect; some cases are felony offenses under UK and U.S. animal protection laws. Many instances of animal cruelty and neglect are indicative of in-family violence and psychopathology, especially empathy-deficit disorder. Involvement in animal welfare and protection clearly crosses over into the One Health consideration of mental health, where good relationships with other species contribute significantly to human physical, mental, and spiritual well-being.

In natural ecosystems there are individual wills to live that are co-evolving in the biodiverse dynamics of ecological homeostasis and health. Johann Wolfgang (von) Goethe (1749–1832) stated, 'Every creature is its own reason to be. All its parts have a direct effect on one another, a relationship to one another, thereby constantly renewing the circle of life; thus, we are justified in considering every

FIGURE 11.4 Veterinary intervention can make an enormous difference for individual animals: Like this neglected puppy, for instance. Rescuing and healing this village pup in India gave me the rewarding gift of his attentive gaze.

animal physiologically perfect. Viewed from within, no part of the animal – as so often thought – is a useless or arbitrary product of the formative impulse' (12).

Imposing the anthropocentric right to live for some species over others, including our own, has contributed to the planetary dysbiosis that we face today. A One Health approach can help override such cultural and political ideologies of selective speciesism.

As animals have served and benefited us for millennia and continue to do so in myriad ways, so we benefit the more we serve and help them as our wards, companions, healers, teachers, patients, and friends. The bond that people have with the animals in their lives must become a boundless circle of compassion, expanding to encompass all living beings if we are to justify keeping any animal as a domesticated companion beyond our selfish needs. Anthropocentrism is a state of mind, culturally condoned, that has yet to be overcome (13).

Animal cruelty can impact the entire family, and investigation, intervention, and prosecution, along with seizing and caring for abused and neglected animals, take time, money, and expertise, including forensic veterinary input. I am heartened that the State of Minnesota in 2024 has Senate and House bills (SF 3542 and HF 3799) under consideration with rising public support for establishing an Office of Animal Protection to be housed under the Bureau of Criminal Apprehension (BCA) of the Department of Public Safety (for details visit MNOAP.org). I wish all states and nations would follow this initiative, since witnessing animal cruelty harms children and can be indicative of in-family – especially spousal – violence. The analysis of 27 studies on domestic abuse and 15 papers on sex offenders identified four dominant themes:

1. Animal abuse as an indicator of domestic abuse
2. Childhood animal abuse as a signal of other problems within a household
3. Animal abuse as an indicator of more severe domestic abuse
4. The inclusion of pet abuse as a unique form of domestic abuse

Both literature compilations highlighted gaps in risk assessment, risk management, and the lack of identification and training in this neglected area to provide prevention and response services (14).

FIGURE 11.5 A declaration on a barn in the Lake District, England, that encapsulates the duty of care.

I call on all municipalities and those concerned about animal, child, and elder abuse and neglect and sexual violence to visit the National Link Coalition (nationallinkcoalition.org), a non-profit, multidisciplinary collaborative initiative to increase awareness and address public policy, programs, and research. This national resource center on the link between animal abuse and human violence seeks to stop the cycle of violence that often affects multiple family members and can spill over to violence in schools, which is on the increase today.

Professor of law and environmental justice, Angus Nurse, in his Research Comment 'Animal Rights and the Language of Human Rights' asks how the veterinary and associated professions dealing with animals should respond to existing and future legal rights and advocates the setting out of a legal framework for animal rights (15).

While domesticated animals have long been regarded as property and wild animals are in a different class, the categorical imperative of treating animals humanely has been a hallmark of civilized society for decades. But now there is a deeper understanding of animal sentience, which could be the basis for setting a legal framework for animal rights, especially through the sciences of cognitive ethology and neurochemistry.

Many animal species, especially those living in social groups, evidence individual differences in temperament. Such differences in humans we regard as individual personalities. Yet to ascribe personalities to non-humans has been considered unscientific anthropomorphizing. This taboo has helped ensure non-human animals will never be accorded legal personhood regardless of their ethological and neurochemical claims! Only a few centuries ago, children and wives were denied legal personhood, along with human slaves.

Now there is a paradigm shift, as Prof. Nurse is indicating, where the veterinary profession, allied animal industries, and animal caregivers have a collective duty to examine how animals are being treated in light of recent advances in animal sentience research. For instance, identification of different personalities in animals and their responsiveness to different conspecifics (16) and African elephants using proper names in specific sounds to address their fellows, an ability thought to be uniquely human (17), raise the ethical and legal issue of animal personhood.

RISING INFLATION AND A NON-SUSTAINABLE ECONOMY

The news reports began around 2023 in the United States and the UK that animal shelters were filling up with relinquished companion animals, especially dogs, because people could no longer afford to keep them and because of housing restrictions. The situation a year later is now much worse. The rising cost of living, of food, housing, health care, utilities, and transportation are a reality, as is the deteriorating condition of essential infrastructures and supply chains.

A failing economy that is non-sustainable for multiple reasons, exacerbated by climate change, is behind public concerns globally about inflation. This is a prelude, if not corrected, to growing poverty, materially and spiritually; economic, political, and environmental refugees; famine, plague, and pestilence; terrorism; insanity; and war.

All of this can still be averted and corrected by each one of us striving to live in accord with the Golden Rule and speak truth to power and corruption and to take Mark Twain's advice to change how leaders lead and support a United Nations of Environmental and Social Justice.

Leonardo da Vinci asserted, 'The time will come when cruelty to animals will be equally punished as cruelty to humans.' This is a call for *equalitarianism*, the ethical bedrock of any eco-democracy. I see this as our only hope to prevent the extinction of all that made and makes us human. The best qualities of human nature, of character, empathy, compassion, respect for life, innate potential, and individual genius, can be best awakened and nurtured in children by appropriate relationships with other species – wild and domesticated – that engender respect and understanding, along with experiences in nature that inspire awe and wonder. The education and inspiration of children will be addressed in more detail in the next and final chapter.

REFERENCES

1. Berry, T. *The Dream of the Earth*. San Francisco, CA, Sierra Books, 1988. See also www.ecozoicstudies.org
2. Fox, M.W. Veterinary bioethics: Ecoveterinary and ethnoveterinary perspectives. *Veterinary Research Communications*, 1995; 19: 9–16 and Fox, M.W. Principles of veterinary bioethics. *Journal of the American Veterinary Medical Association*, 2006; 229: 666–667
3. Illich, I. *Medical Nemesis: The Expropriation of Health*. New York, Bantam, 1977.
4. Lans, C. Validation of ethnoveterinary medicinal treatments. *Veterinary Parasitology*, 2011; 178(3–4): 389–390.
5. Lans, C. et al. Ethnoveterinary medicine: Potential solutions for large-scale problems. In Susan G. Wynn and Barbara J. Fougere (eds.), *Veterinary Herbal Medicine* (pp. 17–32). St. Louis, MO, Mosby Elsevier, 2007.
6. Mathias, E. Ethnoveterinary medicine: Harnessing its potential. *Veterinary Bulletin*, 2004; 74(8): 27N–37N.
7. McCulloch, S. Vets must be strong advocates for animal rights. *Veterinary Record*, 2023; 192(12): 491. https://doi.org/10.1002/vetr.3205
8. British Veterinary Association. *UK Sustainable Animal Agriculture*. https://www.bva.co.uk/take-action/our-policies/uk-sustainable-animal-agriculture/#) and Wensley, S. *Less and better – A call to value our food*. April 23, 2019. https://www.bva.co.uk/news-and-blog/blog-article/less-and-better-a-call-to-value-our-food/
9. Trouwborst, A. et al. Domestic cats and their impacts on biodiversity: A blind spot in the application of nature conservation law. *People and Nature*, 2020; 2: 235–250. https://doi.org/10.1002/pan3.10073

10. Tan S.M.L. et al. Uncontrolled outdoor access for cats: An assessment of risks and benefits. *Animals (Basel)*, 2020 Feb 6; 10(2): 258.
11. Rodríguez-Rodríguez, E. et al. Feral animal populations: Separating threats from opportunities. 2022. https://doi.org/10.13140/RG.2.2.21526.22088/1
12. Miller, D. *Goethe: Scientific Studies.* Princeton, Princeton University Press, 1995.
13. Lockwood, R. Anthropomorphism and anthropocentrism in the Anthropocene. In Fine, A. et al. (eds.), *The Handbook on Human Animal Interactions and Anthrozoology.* London, Routledge Press, 2023.
14. Diemer, K. et al. Exploring the linkages between animal abuse, domestic abuse, and sexual offending: A scoping review. *Health & Social Care in the Community*, 2024; 1170505.
15. Nurse, A. Animal rights and the language of human rights. *Veterinary Record*, 2024; 194: 389–390.
16. Vonk, J., and Eaton, T. Personality in nonhuman animals: Comparative perspectives and applications. In V. Zeigler-Hill and T.K. Shackelford (eds.), *The SAGE Handbook of Personality and Individual Differences: The Science of Personality and Individual Differences* (pp. 23–51). Sage, 2018. https://doi.org/10.4135/9781526451163.n2
17. Pardo, M.A. et al. African elephants address one another with individually specific name-like calls. *Nature Ecology & Evolution*, 2024; 8: 1353–1364. https://doi.org/10.1038/s41559-024-02420-w

12 Humane and Environmental Education

No Child Uninformed

In their relevant article, 'The Development of Speciesism: Age-Related Differences in the Moral View of Animals,' the British researchers summarized their findings as follows:

> Humans care for the well-being of some animals (e.g., dogs) yet tacitly endorse the maltreatment of others (e.g., pigs). What treatment is deemed morally appropriate for an animal can depend on whether the animal is characterized as 'food.' When such categorization of animals emerges and when a moral hierarchy of beings depending on their species membership (speciesism) develops is poorly understood. We investigate this development across samples of children (9–11 years old), young adults (18–21 years old), and adults (29–59 years old; total N = 479). Compared with young adults and adults, children (a) show less speciesism, (b) are less likely to categorize farm animals as food than pets, (c) think farm animals ought to be treated better, and (d) deem eating meat and animal products to be less morally acceptable. These findings imply that there are key age-related differences in our moral view of an [animal's] worth that point to socially constructed development over the lifespan. (1)

Evidently, the missing link between the fully human and other animals has been discovered. It is the most carnivorous, invasive, cruel, and murderous of all the primate species. It is us; we are this missing link, as my friend the late ethologist, Konrad Lorenz, would often proclaim! The chimeric, bipolar nature of this modern protohuman psyche, from rescuer to rapist, healer to harmer, and artist to autocrat, is a product of familial and cultural epigenetics, for better and for worse.

The best of human nature can be nurtured and inspired by humane education and example, striving to live in accord with the Golden Rule and the virtues extended to embrace respect for all life. Planting the seeds of compassion and empathy in childhood, facilitated by living with an animal companion like a loyal and loving dog, can do much to help us evolve into a more civilized species. Cruelty toward animals in childhood has been linked to sociopathic and psychopathic violence in adulthood.

DOI: 10.1201/9781003569787-12

FIGURE 12.1 The author with his rabbit 'Thumper,' who was his trusting friend and playmate.

Children who grow up in a culture of violence toward nature and other species need help learning how they can make a difference and not become desensitized and accepting of cultural 'norms' of inhumanity, speciesism, and racism. Several universities are now offering courses in humane education, with one – Antioch University in partnership with the Institute for Humane Education – offering online MEd, MA, Ed, and graduate certificate programs specializing in humane education. For details visit: https://www.antioch.edu/academics/education/humane-education-ma/. The creator of these programs, Zoe Weil, is also author of *The World Becomes What We Teach* (2).

To realize that we are all part of the cosmic miracle of life and consciousness puts the significance of our individual existence in the broader context of awakening our sense of kinship with all life. This step is necessary to transcend self-centeredness and anthropocentrism. Empathic sensitivity and ethical sensibility may then arise spontaneously, reducing the need for moral instruction and law enforcement.

Corporate as well as parental and institutional responsibility are too often lacking. For instance, new research sheds light on the scope of the livestock industry's influence over prominent agricultural research centers at two public universities. The article, published in the journal *Climatic Change*, details how the meat industry funds credentialed academics to 'obstruct unfavorable policies,' especially those targeting the industry's largely unregulated methane emissions (3). The Rachel Carson Campus Network is one of several non-profit organizations reaching out to college students to inform, empower, and promote environmental justice advocacy. Land grant universities are notoriously beholden to agricultural interests at both educational and graduate research levels.

There needs to be greater vigilance at all educational levels – from kindergarten to college; technical and trade schools; and graduate schools in science, economics, medicine, and veterinary medicine – over corporate influences on the curricula and the open exchange of information. The strong alliance between the pharmaceutical and agribusiness food industries and federal and state regulatory agencies with mainstream professional organizations like the American Medical Association and American Veterinary Medical Association, and their collective influence on medical and veterinary student education, need to be addressed and assessed from an objective, bioethical perspective – especially in terms of humane and environmental responsibilities.

No child should be left behind or disinformed when it comes to humane and environmental education.

FIGURE 12.2 All children benefit from animal companionship, enjoying the communion of greeting and playing together when there is trust and understanding.

Impaired sensitivity toward animals may or may not be caused by a child being deprived of meaningful contact with animals, since other factors are involved in the genesis of what I call animal insensitivity syndrome. It is a cognitive and affective developmental disorder that I see as part of a larger problem of insensitivity and indifference toward the environment.

I was heartened to learn that Richard Louv's book, *Last Child in the Woods: Saving Our Children from Nature Deficit Disorder* (4), has been widely embraced by educators and parents. In the common currency of compassion and respect, our transactions and relationships with each other, with other animals, and with the Earth or natural world are ideally framed within the Golden Rule of treating others as we would have them treat us. This currency includes such ancient coins of wisdom as altruism (enlightened selfishness), *ahimsa* (Sanskrit for not harming in any way), and *karma* (having prescience and understanding that what goes around, comes around). All our choices and actions have consequences.

Ethical blindness that comes from a lack of empathy with other living systems and beings is linked with a lack of respect and understanding that when we harm animals and the Earth, we harm ourselves – especially in the production of food and fiber, and indirectly through our dietary choices, consumer habits, and lifestyles.

We harm animals by destroying their natural habitats and in making them suffer so that we may find new and profitable ways to cure our many diseases. The actual prevention of disease is in another domain based on an entirely different currency from what is still the norm in these sickening times. The currency of unbridled exploitation and destruction of natural resources and ecosystems and the wholesale commercial exploitation of animals cannot continue because it is not sustainable. One of the greatest sicknesses is the proliferation of factory livestock farms – the intensive confinement systems that are stressful to the animals, promote disease, are environmentally damaging, are ecologically non-sustainable, and put consumers at risk.

Sustainable rates of exchange are based in part on mutual aid, a point emphasized by Russian Count Peter Kropotkin, who envisioned the ideal human community like a functioning ecosystem of interdependent, democratically integrated individuals and species creating mutually enhancing, symbiotic, micro and macro communities that he discovered in his studies of the co-evolved flora and fauna of the vast wild steppes of his native land (5). Finding no dictatorial hierarchy in nature, he coined the term 'anarchy' in his book *Mutual Aid*, first published in 1902. Since this term now has negative societal associations, I would instead use the term 'holarchy.'

Nature deficit disorder leads ultimately to regarding and treating the living Earth as a non-living resource, just as animal insensitivity syndrome can lead to animals being treated as mere objects. Insensitive, indifferent, and cruel contact and experiences with animals during the early years, probably a critical sensitization/desensitization developmental stage, can mean a poorly developed and extremely self-limiting capacity to empathize with others; to be able to recognize, anticipate, experience, and share other's feelings; and to express and deeply consider one's own. Adult rationalization, denial, and ethical blindness are rooted in early childhood conditioning and desensitization, leading to acceptance and eventual participation in many forms of animal exploitation and cruelty. This is vividly documented by British hunt saboteur, Mike Huskisson, reporting children witnessing deer and fox hunting and being ritualistically 'bloodied' and receiving parts of the murdered animals to take home either to eat or as prized trophies – mementoes of their presence at the kill (6).

In some countries such as India where I have worked with my wife, Deanna Krantz, we have both witnessed how people simply turn a blind eye to the suffering animal and the polluted stream because they themselves are struggling to survive. Individuals who feel helpless become desensitized, resigned fatalists. In some contexts, intervention to help a suffering animal or to stop a stream from being poisoned by a tannery or a slaughterhouse could mean death threats and violence. Yet when communities learned we provided free veterinary care and shelter for injured and sick animals, people who would formerly ignore an animal's plight would come from miles around, often accompanied by their children, with animals for us to care for or to adopt a healthy one from us (7). Most people are not indifferent where there is hope and recourse.

Observing another's suffering and being unable to do anything to help leads to learned helplessness. Seeing others suffering and being indifferent about it is the next step toward the total disconnect of empathy, termed bystander apathy. The next step is to observe and derive vicarious pleasure in witnessing another's plight (8). This is but one small step away from deliberate torture and calculated cruelty, either perpetrated alone or in participation with others, as in the name of entertainment, sport, quasi-religious or cult ritual, and (as some see it) experimental vivisection.

Animal suffering matters because it is a matter of conscience. Deliberate cruelty toward animals and acceptance and indifference toward their plight is unconscionable – a zoopathic state of mind. This parallels the behavior and cognitive and affective impairment of the sociopath and of the 'ecopath,' who has no twinge of conscience over the destruction of the natural environment. Where there is a lack of empathy, of feeling for others, there can be neither concern nor conscience.

Becoming desensitized to animal suffering and then treating animals as mere things, as objects devoid of sentience, is part of the same currency as treating fellow humans as objects rather than as subjects. Such dehumanization, coupled with demonization, can lead to genocide and, more commonly, to ecocide and 'speciesicide.' This is the annihilation of animal species and their communities that are perceived as a threat or of no worth. Our attitudes toward animals, the degree of ethical concern and moral consideration, can mirror our regard for each other – for better or for worse. When collectively our hearts and minds are open to the tragedy of reality and we really see and feel all that is going on around us, empathizing fully with another's suffering, times will begin to change for the better. Young people becoming informed and engaged in relevant animal and environmental protection issues are empowered and enabled to work together to creatively redirect what the American Psychological Association terms eco-anxiety.

This is seen as a threat by some sectors – notably by India's Prime Minister Narendra Modi, whose purging of the theory of evolution from school textbooks in 2023 adds to an already long list of subjects expunged from education. After he had given a talk in Delhi in 2014 to students that was televised under mandate to all schools across India, he stated in response to a student's concern about climate change that 'The climate has not changed. We have changed.'

Social justice and environmental justice are essential elements of a viable democracy. Likewise, human and non-human rights should be given equal and fair consideration in any civil society.

EXAMINING OUR ANIMAL RELATIONSHIPS

Children may wonder and ask why most wild animals flee when they see us and are obviously fearful. We should be honest and tell them that it is an instinctual reaction because we humans, and our prehuman primate ancestors before us, have been killing animals or driving them away for close to 3 million years.

Those animals taken from the wild, such as wild horses and elephants, have their spirits broken before they will serve us. But many animal species when taken soon after they are born will become emotionally attached to us, dependent and trusting. Such trust enabled our ancestors to begin to domesticate them for various purposes, beginning about 10,000 years ago with dogs, and later, sheep, cattle, and horses. In many instances and relationships, we betrayed that trust.

All these human purposes and our relationships with animals, including wildlife being 'harvested' by hunters, trappers, and fishers, need to be examined by all who feel affection and concern for animals and enjoy being out in nature, which would not exist without the animals. Nor would we. Fortunately, there are individuals and organizations dedicated to improving the care and welfare of animals, domesticated and wild, and advocating for their rights and interests, protection, and conservation.

Gratitude toward all creatures great and small will enable us to break away from our anthropocentric, ancestral past as animal exploiters and killers – a carnivorous

FIGURE 12.3 A coyote in a leg-hold trap, cut down to the bone.

primate species. This will enable the blossoming of our humanity, of compassion, empathy, and respect for all living beings.

ANTIDOTES AND SPONTANEOUS EMPATHY

The antidotes are many. Those in Richard Louv's book should be coupled with meaningful contact with companion and other animals, with parental supervision and humane instruction to foster respect, self-restraint, gentleness, patient observation, and understanding.

I published the following letter in my syndicated newspaper column *Animal Doctor* (July 11, 2017) from a mother whose son demonstrated empathic concern for animals at an early age:

> Dear Dr. Fox,
>
> I read your recent column about the need to bring more environmental and humane education into schools and applaud the efforts of your daughter Camilla with

Project Coyote, and others who are doing this. I work for Pro Animale in Germany whose book Memento you recently reviewed, and I want to share this about my son's natural empathy toward animals which was quite independent of any influences from me during his early years.

Since he started talking when he was only 2 years old I could tell that he has had a very special sense and empathy for animals and their feelings and needs. When he once saw a just hatched chick in an incubator at a fair, he did not say "how cute", he immediately looked concerned and said it looks for its mother we need to find her. In a wildlife park he saw two raccoons in an enclosure sitting by the fence and said they look sad and do not want to be fenced in but free. When we were on a boat ride to see seals on a sandbank the tour operators suddenly let a net down on the bottom of the sea and fished out crabs and other creatures from there to show to the people and tell them about them. Yukon, 3 years old then, got really mad at the seamen and told them to let the crabs immediately back into the water, because they do not want to be taken out. Unfortunately, we were not heard by the seamen even when I said that it is very harmful in general for the creatures on the bottom of the sea when they let a net down every time, they have a boat ride to the sandbanks. They did not understand because they thought since they put the creatures they fished out in a tank with water and later back into the sea, that it was okay. When one person said to Yukon "look at these interesting animals in the tank," Yukon responded "I look at the animals who are in the sea where they want to be" and looked down into the water where he couldn't see any animals but I think he imagined seeing them there. I was so amazed how a three-year-old can be more understanding of animals and their needs than grown up people. I am sure you can tell I am very proud of him.

–S.B., Sennfeld, Bavaria, Germany

DEAR S.B.,

I am sure that many parents reading your letter will have had similar experiences with their young children's reactions to animals affirming my contention that they have a natural affinity for fellow creatures which is the foundation for empathy and compassion as they mature. But cultural norms regarding accepted treatments of animals and the attitudes and reactions of adults can either facilitate or inhibit the development of empathy, which some regard as being a sissy and not facing reality. Indeed, empathy can be a burden especially when not supported by others and when not expressed in appropriate action or choice.

Your son reminds me of my younger daughter Mara who, around the same age as Yukon, said that she was going to friends at Thanksgiving to eat turkey. Her stepmother and I asked her if she knew what the word turkey meant. Since she did not know, we gently told her that "turkey" was a bird. Her immediate response was one of shock and she exclaimed that she would never eat a bird or any animal. She later told us that at the gathering she refused to eat any turkey but was told that it was OK because the farmer had killed the bird. From that time on, with no prompting from us, she decided to become a vegetarian. My other two children Camilla and Mike Jr. decided at a later age to become vegan for ethical, humane, and environmental reasons.

> In sum, all children, with rare exception, have the capacity to identify with and show empathy toward others, the absence of which has been linked to a lack of conscience, feeling for others, dissociation, and sociopathic behaviors in later life. It is an attribute best guided by example and enabling the child to make informed choices and to share and question openly, without ridicule, how they feel about animals and how they think they should be treated. For more details see the seminal book by my former graduate student Dr. Randall Lockwood, co-authored with Frank K. Ascione. *Cruelty to Animals and Interpersonal Violence* published by Purdue University Press. 1998.

A child's sense of wonder, if it is nurtured and not crushed or left to wither, blossoms into the adult sense of the sacred – an ethical sensibility of respect for the sanctity of every life. Such sensibility is invaluable in difficult times such as these, where we see so much abuse of life and irreverent treatment of other beings with whom we share the kinship of sentience and of genetic ancestry.

FIGURE 12.4 The author having an early lesson in animal behavior from a friendly dog.

A child's sense of curiosity leads to natural science and instrumental knowledge. Combined with a sense of wonder, curiosity leads to imagination and creativity, while the sense of the sacred is the foundation for an ethical and just society and empathetic, caring, and fulfilling relationships – human and non-human. This empathy-based bioethical and moral sensibility that gives equally fair consideration to all members of the biotic community is an ideal that may yet become a reality, provided the potential for such development is properly nurtured and reinforced by example in early childhood.

Young children must be informed as to why it is in everyone's best interests to care for all animals and the environment and what they can do to help. Dr. Jane Goodall has set up environmental and humane education programs (https://www.rootsandshoots.org) in many schools around the world, as have others concerned about animals and the natural environment. These include the National Humane Education Society (https://nhes.org/education-2); my daughter Camilla Fox's Project Coyote (http://www.projectcoyote.org/programs/keeping-it-wild/), and another of my former graduate students, Prof. Emeritus Marc Bekoff, a prolific author and advocate, as per his book *Rewilding Our Hearts: Building Pathways of Compassion and Coexistence.*

All educators need to engage young minds with these concerns (involvement is evolvement) and sow seeds of hope for their own good and for the revolution in planetary consciousness, conscience, and conservation. Wisdom speaks when we listen with our hearts, as when we enter a forest, and the forest enters us and when we empathize with other creatures great and small.

Children will be enthralled to discover how so many animals, plants, and microorganisms benefit each other and us, from pollinating honeybees and rain-gathering forests to the good bacteria in the soil and in our guts, without whom we could not live. There is a wealth of educational materials to help children understand how similar and often identical our basic needs and feelings are to those of many other animal species, notably how they too can show fear and experience anxiety, anticipation, and grief. Animals are also able to exhibit empathy and altruism; express pleasure in social greeting and caregiving; and express joy, self-restraint, and creative imagination while engaging in play. Many have physical and sensory attributes from speed, strength, and survival skills to sonar echolocation and celestial navigation abilities, which put a big question on the notion of human 'superiority.' Such knowledge can deepen respect and appreciation for other living beings, especially when coupled with learning how animals have benefitted humanity in so many ways since the beginning of recorded history and continue to do so today as they act as our companions, teachers, and healers.

This brings into question human choices, ethical decision making, and how best to direct our relative freedom of will and impulses to avoid harming others and

FIGURE 12.5 Teaching children through books and video documentation about how other animals communicate and care for their young and each other, which can do much to break the cycles of animal cruelty, neglect, indifference, and recreational 'sport' hunting and trapping.

unwittingly playing the role of predators, parasites, and invasive and infectious organisms. Such analogous behaviors we see on occasion in our own species in those still ruled by anthropocentrism and misled by chauvinistic and other prejudicial, even genocidal, ideologies and societies. We can also examine our own assumed rights, entitlements, and cultural traditions against those within our communities and beyond. Is there not more conflict arising from the lack of family planning around the world, causing a rising population with an insatiable demand for meat, fish, and other foods from animals on the one hand, while millions of others are landless and starving?

To shield children from such realities 'until they are old enough to understand' is a form of denial and obfuscation that I have confronted while giving talks to schoolchildren who, from middle school on, should be encouraged and guided into this realm of ethical human existence and ultimate responsibility.

Like our physical health, our mental health and Earth health are deeply interconnected. For us to be whole in body, mind, and spirit, we not only need our connections with each other in our communities and cultures but also beyond with animals and the Earth. These extended connections must be properly established in early childhood to prevent the harmful consequences of nature and empathy deficit disorders and animal insensitivity syndrome.

Empathy is the bridge to self-transcendence, as exemplified by the 'I–thou' relationship described by philosopher Martin Buber. This is comparable in some respects to the Christian's *agape* and the mystic's ecstatic rapture in the feeling of unity and oneness. In their book, *In Search of Self: Interdisciplinary Perspectives on Personhood*, J. Wenzel van Huyussteen and Erik P. Weibe state that 'it is through the transcendence of empathy that one gains the ability to separate self from other and to see the other as fully other in relation to the self. Through the transcendence of imagination, one receives release from the past through openness to a new future' (9). There now is considerable literature to further the exploration and understanding of empathy and the growing recognition of its role as an essential element in child rearing and education. It also has evolutionary significance in us and other species and contributes to healing, to the arts, and has an emerging potential to transform dystopian societies into sane and humane Earth communities (10–12). Social psychologist Jonathan Haidt, in his book, *The Anxious Generation* (13), documents the rise in anxiety, depression, self-mutilation, violence, and suicide in children and young adults who have increased involvement in social media. Rather than stare at smartphones and Instagram, young people need childhoods based more on time-outs and play. But as I emphasized in my book, *Bringing Life to Ethics: Global Bioethics for a Humane Society* (14), such transformation will be illusory if our collective empathy is limited by anthropocentrism and does not embrace *all* sentient beings and the environment which we of the Earth community share. Children need instructive, supervised time playing and exploring the natural world that can help put their lives in perspective with the realization that they are part of the miracle of existence and that no life is without purpose, including theirs. Our ultimate well-being is in that freedom of spirit which enables and empowers feeling for, loving, and living for others.

When I contemplate suicide, I think of all the suffering it would cause to all who love me and need me and whom I love, including my dog and cat. I wonder how many animal companions have given children and adults something/someone to live for. A good dog in a child's home might prevent the mental health problems documented by Dr. Haidt, given parental instruction on how to properly care for the non-human family member. Dogs in particular play a significant role in children developing a strong immune system and related healthy gut microbiome. Dogs also help increase children's physical activity, helping to reduce childhood obesity and sedentarism (15).

Having an animal companion in the home is likely to make children more responsible citizens and caring parents. As the Australian aborigines proclaim, 'Dingo (dog) make us human.' It would be a different world if they grew up to have the empathy of Australian aboriginal Bill Neidjie, who said, 'If you feel sore – headache, sore body that means somebody killing tree or grass. You feel because your body is in that tree or earth. Nobody can tell you, you got to feel it yourself' (16). There is music in the forests, bamboo jungle, grasslands, running streams, and moon-drawn oceans. Much of our musical sensibility and instruments are derivative of these sources, as evidenced by indigenous musicians around the world, along with their dances and songs, often mimicking animals. Imagine children choosing, even making, their own musical instruments and accompanied outdoors listening to the night frogs and insects, or to the birds at dawn, and responding spontaneously with their instruments! It would likely be like a 'first communion,' as the children let their spirits fly, inspiring the spirits of others. This type of social bonding (17) would be a much healthier activity than scrolling through social media! Thousands of children and adults have enjoyed gaming with wolves in the educational video game WolfQuest, enabling participants to learn about wolves and interactively experience what it is like to be a wolf.

I wish that all adolescents were able to engage in the equivalent of the Native American Indian vision quest and that educational institutions would open their doors and minds to some of the traditional wisdom of indigenous peoples in caring for younger generations and helping them mature into responsible adulthood. Climate change is exacerbating mental disorders. A global survey in 2021 found that more than half of people aged 16–25 felt sad, anxious, or powerless or had other negative emotions about climate change. Altogether, hundreds of millions of people might be experiencing some type of negative psychological response to the climate crisis. Fifty-eight percent felt betrayed by their government (18). But on the flip side, studies suggest that individuals, by becoming informed and involved with others in taking action to combat climate change, can temper their eco-anxiety. To this end, all schools should have environmental education courses and updated home economics courses to enable students to better cope and adapt to conditions that are likely to challenge them sooner rather than later in life.

I will close this chapter with some relevant lines from poet Walt Whitman (1819–1892):

> There was a child went forth every day,
>
> And the first object he looked upon and received with wonder or pity or love or dread, that object he became,
>
> And that object became part of him for the day or a certain part of the day … or for many years or stretching cycles of years.

REFERENCES

1. McGuire, L. et al. The development of speciesism: Age-related differences in the moral view of animals. *Social Psychological and Personality Science*, 2023; 14(2): 228–237. https://doi.org/10.1177/19485506221086182
2. Weil, Z. *The World Becomes What We Teach*. Newburyport, MA, Lanter Press and Media, 2021.

3. Morris, V. and Jacquet, J. The animal agriculture industry, US universities, and the obstruction of climate understanding and policy. *Climatic Change*, 2024; 177: 41. https://doi.org/10.1007/s10584-024-03690-w
4. Louv, R. *Last Child in the Woods: Saving Our Children from Nature Deficit Disorder.* New York, Algonquin Books, 2005.
5. Kropotkin, P. *Mutual Aid: An Illuminated Factor on Evolution.* Oakland, CA, PM Press, 2021.
6. Huskisson, M. *Outfoxed Again: Winning for Animals.* Animal Welfare Information Service. Halesworth, Suffolk, UK. 2017.
7. Krantz, D. and Fox, M.W. *India's Animals: Helping the Sacred and the Suffering.* Wilmington, DE, Mozart & Reason, Wolfe Ltd., 2016.
8. Fox M.W. Ways of seeing: Animals in art and life. In *Animals and Nature First: Creating New Covenants with Animals and Nature* (pp. 168–174). CreateSpace Publishing/Amazon.com, 2011.
9. Wenzel van Huyussteen, J. and Wiebe, E.P. *In Search of Self: Interdisciplinary Perspectives on Personhood.* Grand Rapids, MI, Eerdmans, 2011.
10. Brener, M. E. *Evolution and Empathy: The Genetic Factor in the Rise of Humanism.* Jefferson, NC, McFarland Press, 2008.
11. Krznaric, R. *Empathy: Why it Matters, and How to Get it.* New York, Penguin Random House, 2014
12. Rifkin, J. *The Empathic Civilization: The Race to Global Consciousness in a World in Crisis.* New York, Tarcher Perigree, 2009.
13. Haidt, J. *The Anxious Generation: How the Great Rewiring of Childhood Is Causing an Epidemic of Mental Illness.* New York, Barnes & Noble, 2024.
14. Fox, MW. *Bringing Life to Ethics: Global Bioethics for a Humane Society.* Albany, NY, State University of New York Press, 2001.
15. Chase, C. J. et al. The KID study (kids interacting with dogs): Piloting a novel approach for measuring dog-facilitated youth physical activity. *Journal for the Measurement of Physical Behaviour*, 2024; 7(1): jmpb.2023-0014.
16. Neidjie, B. Statement. In *The Boundless Circle: Caring for Creatures and Creation* (Ch. 7, p. 144). Wheaton, IL, Quest Books, 1996.
17. Savage, P.E. et al. Music as a coevolved system for social bonding. *Behavioral and Brain Sciences*, 2021; 44: 1–22. https://centaur.reading.ac.uk/95527/
18. Hickman, C. et al. Climate anxiety in children and young people and their beliefs about government responses to climate change: a global survey. *Lancet Planetary Health*, 2021 Dec; 5(12): e863–e873. https://doi.org/10.1016/S2542-5196(21)00278-3

Postscript

For All My Relations – A Brief Autobiography

My quality time with canine companions as a child led me to become a veterinarian so I could help all animals and heal what I witnessed to be a broken bond between my own kind and animal kind – all our relations. This concern took me along the scientific path of researching animal behavior/ethology, sentience, and from there to animal rights and human responsibilities, including our collective responsibility for the natural environment upon which the health of all life depends. There will be progress once the integrative concept of One Earth and One Health is realized and put into practice and public policy.

I grew up during World War II in the industrial, cotton milling, and coal mining county of Lancashire in the north of England, surrounded by William Blake's 'satanic mills.' I came to respect the dignity of hard labor, the spirit of the Luddite, and the virtues of frugality, self-reliance, and serving the common good in accord with the Golden Rule. I also witnessed how humans can harm the natural environment and create toxic wastelands. Animals taught me through their trust and affection to be gentle and respectful, qualities of character that were lacking in some of my peers, who had blown up frogs with straws at one of the local ponds where I played and studied life in the waters, and even some adults, who had thrown a sack in the pond which I opened and found a litter of drowned kittens.

My father was abroad in World War II, so my mother cared for stray and injured animals. Walking home from school one day at 8 years of age, something impelled me to I look though the yard fence of the veterinary hospital as I passed it. I saw two large trash cans abuzz with flies, brimming over with dead dogs, cats, puppies, and kittens. This tragic sight was incomprehensible to me, and I realized that there was something seriously wrong and resolved to become a veterinarian to help save animals. This mass 'mercy killing,' I learned later, was one of the dark consequences of food shortages during World War II.

A year or so later I went racing through our neighborhood to catch our rescued dog, 'Rover,' who was having a 'running fit,' snapping at the air and in obvious, hysterical terror. He had stolen and eaten a loaf of white bread delivered by the local baker, who had set it on our immediate neighbor's porch. My father was a health food advocate and only allowed whole wheat bread into our home.

DOI: 10.1201/9781003569787-13

I later learned that dogs developed this kind of hysteria after consuming 'agenized' bread, which is bleached with nitrogen trichloride, which produces the neurotoxin methionine sulfoximine. This experience made me aware of the risks of processed and adulterated foods.

I was fortunate to attend a small-town high school, Buxton College, where I became troop leader of the school's Boy Scouts and won the Queen's Scout award (U.S. Eagle Scout equivalent). My extracurricular training in elocution, which later served me well in public speaking, enabled me to execute a well-received title role of Hamlet in the town's Opera House theatre, which accommodated the school's annual production of Shakespeare's play. Several years of fencing in the Amateur Fencing League lent additional sparkle to my stage performances!

My childhood interest in natural history and the life in ponds and streams lead me to focus on the *Trichoptera* (caddis fly), and before graduation from high school, I was elected a fellow of the Royal Entomological Society for my studies of these insects. I only later questioned the cultural norms at that time of killing animals in the name of science and 'medical progress.' Thanks to these engagements with animals and nature, I was encouraged by my parents and shadowed the local country vet in my early teens when we moved to the sheep-raising county of Derbyshire. This county, famed for its dales and moors, which I regularly hiked with my dog, provided further inspiration to become a veterinarian. My first task at the veterinary clinic at the age of 15 was to kill a terminally ill kitten with a captive-bolt pistol. I saw that as a test of not simply ability but also of the strength of my commitment and resolve to end suffering regardless of my affection and concern for the poor animal. But most urgent, I soon came to believe, was the need to help restore human–animal relationships, which I felt were dysfunctional at best and cruelly exploitative at worst.

On the summer morning that I received acceptance to the Royal Veterinary College in London, I went for a celebratory hike in my beloved Derbyshire dales and moors, flying with elation and the wind in my back, until I was confronted by the horror of a warren of rabbits all out in the bright sunlight blindly circling, tumbling, and convulsing, heads swollen and eyes bulging, many already dead. I lost count of how many skulls I crushed under my hiking boots to end their suffering. They had myxomatosis, a viral disease allowed to spread by the UK government because most land owners regarded rabbits as pests. Once more I was reminded of the potential for cruel and callous indifference in my own species and the challenge of healing human–non-human relationships.

I entered college in the fall of 1957, earning silver and bronze medal awards in animal husbandry and pathology, respectively, and the gold medal and fellowship of the Royal Veterinary College Medical Association for my 1961 thesis on the nutrition and husbandry of working sheepdogs, which documented nutrient-deficient disease in many of these often inadequately nourished animals. Soon after graduating in 1962, and after a brief time as house surgeon at Cambridge University School of Veterinary Medicine, I came to the United States on a research fellowship provided by Dr. John Paul Scott of the Jackson Memorial Research Laboratory, Bar Harbor, Maine, to study canine behavioral and neurological development. This was in part based on a published study of the neurological development of puppies, which I had sent to him and his associate, Dr. John L. Fuller. Further studies were followed at the animal research facility at the state research hospital in Galesburg, Illinois, which enabled me to complete a University of London, England, external PhD dissertation. This helped legitimize my scientific credibility and efforts to contribute to the advancement of animal welfare and rights by applying the science of ethology and developmental psychobiology to animal care and veterinary education. I joined with others to form the International Society for Veterinary Ethology to call for the inclusion of the science of animal behavior/ethology in the veterinary and animal science teaching curricula to improve how animals are cared for and regarded, be they wild or domesticated, leading to consideration of their needs, interests, well-being, and rights.

My first two books were *Canine Behavior* (1965) and *Canine Pediatrics* (1996). In 1968, I edited and contributed to the first veterinary medical and scientific textbook on this subject, *Abnormal Behavior in Animals* published by W.B. Saunders, and in 1971 the University of Chicago Press published my PhD dissertation, *Integrative Development of Brain and Behavior in the Dog.* Some of my dissertation findings were applied to the U.S. Army's canine biosensor/superdog project and improved the welfare and performance of in-field military dogs in the Vietnam War. My popular books *Understanding Your Dog* (1972) and *Understanding Your Cat* (1974) underscored the value of applying ethology in the care and enjoyment of companion animals.

FROM RESEARCH TO ADVOCACY

Researching the behavior and development of wolves, coyotes, and other wild canids to understand more of the effects of domestication on the dog during my tenured associate professor of psychology position at Washington University, St. Louis, Missouri, where I taught animal behavior/ethology and developmental psychobiology, I published several scientific articles and books, such as *The Behavior of Wolves, Dogs and Related Canids* (1971) and *The Dog: Its Domestication & Behavior* (1978). This research was earned me a DSc degree in ethology/animal behavior from the University of London. A series of lectures at the University of Minnesota, published in *Concepts in Ethology: Animal & Human Behavior* (1974), helped establish the relevance of the science of animal behavior in the use and treatment of animals in biomedical research.

It was a welcome affirmation to receive an invitation from veterinary college dean, Prof. Leo Bustad, to give the graduation address at the School of Veterinary

Medicine at Washington State University in Pullman, Washington, in 1978. I also was greatly encouraged during these several years by Prof. Calvin Schwabe at the College of Veterinary Medicine at Davis, California, who engaged me in giving a series of lectures there, and at the University of Minnesota in the published Wesley W. Spink Lectures on Comparative Medicine, *Concepts in Ethology: Animal and Human Behavior* (1974).

My growing affection and respect for the wolf, detailed in my book *The Soul of the Wolf* (1980), was contrary to the prevailing American view that sought their extermination. I joined others in successfully securing endangered species status to protect the wolf and censoring wild canid eradication methods and consequences. Of several children's books published during this period, three were widely read in schools: *The Wolf* (1973), *Wild Dogs Three* (1977), and *Animals Have Rights Too* (1991). After training and certification as a massage therapist, I contributed to the advancement of holistic, integrative veterinary medicine by writing the book on therapeutic massage for companion animals, *The Healing Touch* (1981), and sequels *The Healing Touch for Cats* (2004) and *The Healing Touch for Dogs* (2004).

Two sabbatical leaves in the Indian jungle studying the dhole, or Asiatic wild dog, gave me a deeper immersion in ecology, conservation, and the politics of extinction, as per my book *The Whistling Hunters: Field Studies of the Asiatic Wild Dog* Cuon alpinus (1984).

I was radicalized by the times – the Vietnam War, the 'War on Wildlife' – especially against wolves and other predators across the United States – and by the ignored realities of climate change associated with a non-sustainable fossil fuel–based economy compounded by petrochemicals, overpopulation, and overconsumption. I was concerned about the family farm crisis and the proliferation of factory farms and puppy mills, which I investigated along with laboratory animal care and zoo facilities. I left academia in 1976 to serve as director of the Institute for the Study of Animal Problems (ISAP), a division of the Humane Society of the United States (HSUS), a non-profit organization in Washington, DC, seeking a broader realm to implement more effective ways to improve the health and well-being of animals. For some years, we published the *International Journal for the Study of Animal Problems* (http://animalstudiesrepository.org/v3_ijsap) and *Advances in Animal Welfare Science*, which helped pave the way for animal welfare science research and bioethics. ISAP helped establish animal welfare as a scientific discipline and relevant field of research. It was eventually replaced by the Center for Respect of Life and Environment (CRLE), which I also directed while concurrently serving as HSUS's Vice President for Farmed Animals & Bioethics. CRLE served as a catalyst to bring various religious and spiritual traditions together to address animal and environmental concerns through various conferences, symposia, and monographs. Many of my writings were later incorporated in the book, *The Boundless Circle: Caring for Creatures & Creation* (1996).

According to the website Activist Facts (https://www.activistfacts.com/person/3365-michael-w-fox/), 'Former HSUS CEO John Hoyt said the Center was created "to let Dr. Fox direct some of his views in a channel that was an arm's length removed from the HSUS." What are those controversial views about which Hoyt was so concerned? Fox holds the radical view that humans and animals have the same moral standing. In his book, *The Inhumane Society*, Fox opines, "The life of an ant and that of my child should be granted equal consideration."

I traveled extensively during these years with the HSUS to document the quality of care and condition of animals in shelters, puppy mills, zoos, laboratory animal research holding and breeding facilities, intensive pig and poultry operations, cattle feedlots, and slaughterhouses. I shared my findings and concerns at various conferences of animal research, farming and animal welfare organizations, and in testimonies on relevant hearings before Congress, state legislators, and governors.

The *Smithsonian* magazine contracted writer David Nevin to interview and accompany me on visits to various animal facilities, including livestock and poultry operations. His article, 'Scientist Helps Stir New Movement for "Animal Rights"', published in April 1980, created such a media firestorm that one Texas senator declared that to be able to generate such media attention, I was being funded by a communist cell bent on overthrowing American agriculture. This was my first encounter with disinformation. I learned that reason was wasted on the unreasonable and that foresight was blinded by denial, ideological rigidity, and lack of insight.

Beginning in the 1980s, I joined many voices from academia calling for acknowledgement of animals' right to equal and fair consideration and for their humane treatment and liberation from all forms of cruel exploitation, coupled with ecological stewardship, which I addressed in a series of books: *Agricide: The Hidden Crisis That Affects Us All* (1986), *Inhumane Society: The American Way of Exploiting Animals* (1990), and *Superpigs and Wondercorn: The Brave New World of Biotechnology – and Where It All May Lead* (1992) [second revised edition entitled *Beyond Evolution: The Genetically Altered Future of Plants, Animals, the Earth – and Humans* (1999) and third updated edition (2004) *Killer Foods: When Scientists Manipulate Genes, Better Is Not Always Best*], and *Eating with Conscience: The Bioethics of Food* (1997).

FACING A RISING OPPOSITION

I was branded as an animal rights radical by the defenders of hunters, trappers, and other exploiters of animals, wild and domesticated, including the U.S. Congressional Sportsmen's Foundation. This was soon after the publication of my book *Returning to Eden: Animal Rights and Human Responsibility* (Viking Press, 1980), which was a mild book compared to my subsequent *Inhumane*

Society: The American Way of Exploiting Animals (St. Martin's Press, 1990). This branding was applied by trapper Russ Carman in his book *The Illusions of Animal Rights* (Krause Publications, 1991), who contended that the movement was one that 'professes the love of animals but is transformed into a hate for humans.' The book *Animal Rights: The Inhumane Crusade* by Daniel T. Oliver, research associate with the Capital Research Center (Merrill Press, 1999), was in part supported by the Koch Foundation. This book presented detailed profiles on the aims and activities of most animal rights and protection organizations in the United States and quotations from some of their spokespersons. Both books essentially drew a dotted line connecting animal rights, vegetarianism, subversion, and terrorism.

My critique of misapplied science and technology, adding to the social ills of humanity, put me in the same league of antisociety/establishment environmental extremists in the eyes of many, including one journalist, Stephen Budiansky. Writing for *U.S. News & World Report* (May 13, 1996), he linked me with Theodore Kaczynski, the so-called Unabomber. After Budiansky's book *The Covenant of the Wild* (1992) was published, in which he presented his thesis that certain opportunistic species deliberately chose to live with humans and thus became domesticated, he was feted by the establishment, even giving a keynote address at an American Veterinary Medical Association conference. This was one of many conferences convened during that time to help keep at bay the concerns of animal rights and liberation advocates and the scientific documentation of ethologists and others concerning the mental states, awareness, and emotionality of animals.

The book *The Case for Animal Experimentation* by another Michael Fox (Canadian philosopher Michael *A.* Fox, not Michael *W.* Fox), published in 1986, caused considerable confusion, consternation, and accusation, since many on both sides of the animal rights issue thought that I had written this book. The 'other' Michael A. Fox subsequently recanted. It is a matter of public record that the opposition to animal rights, animal liberation, environmental ethics, deep ecology, and Earth First! by a host of vested interests in the last quarter of the 20th century morphed into the antiscience, business-first ethos that spread during the first quarter of the 21st century across the United States and abroad.

During this phase of my career, I was walking across the campus to present a lecture to the students at one midwestern veterinary college, when I was intercepted by an embarrassed associate dean who told me apologetically that my lecture had been cancelled and that security would be called if I did not vacate the premises immediately. Just before another lecture to the student body on factory farming and animal rights at the University of Minnesota Rochester, the graduate student organizer informed me that the powers that be had instructed the university bookstore and the main bookstore in the town not to carry any of my books!

So, it was no surprise to me that one of the board members of HSUS, for whom I was serving as scientific director, one Robert Marshak, DVM, dean of the University of Pennsylvania School of Veterinary Medicine, sought at that time to have me fired from HSUS because I was becoming a 'dangerous cult figure,' according to the CEO, John Hoyt. Hoyt retained me after I was called to the University of Pennsylvania campus to defend my concerns with an oral presentation and a written report about the welfare, health, and rights of 'factory' farmed animals before a panel of various department heads involved in the livestock and poultry industries, accompanied by John Hoyt and the HSUS's attorney (who opposed the concept of animal rights).

My report was subsequently given the German Felix-Wankel international award for research into animal welfare and published as a book in 1984: *Farm Animals: Husbandry, Behavior and Veterinary Care*, not long after Dean Marshak resigned from the HSUS board of directors. I followed up this book with a less technical one for consumers, *Eating with Conscience: The Bioethics of Food,* in 1997.

I was unsurprised by the lack of evident responsiveness by this University of Pennsylvania veterinary panel when I raised the issue of animals' social and emotional needs, because earlier, at the inaugural meeting of the Institute for Laboratory Animal Resources Committee on Laboratory Animal Ethology, which I chaired at the National Academy of Sciences in Washington, DC, I had my first encounter with this mindset. The director of the U.S. government's New England Regional Primate Center scoffed and shook his head in dismissal as a non-issue when I raised the question concerning the emotional needs of primates confined in small laboratory cages. In 1986 the State University of New York Press published my scientific treatise, *Laboratory Animal Husbandry: Ethology, Welfare and Experimental Variables*, to legitimize my concerns, documenting how animals' stress and emotional distress were uncontrolled variables when not considered as influencing experimental results and data interpretation.

I helped close a university project contracted for the military on the effects of irradiation on approximately 30 beagle dogs. The dogs were so stressed by prolonged confinement that their condition would invalidate any findings. In another commercial research facility I saw several beagles, the canine equivalent of the standard laboratory rat, living in plexiglass boxes where they were given calculated quantities of air to breathe along with tobacco smoke. I once debated on some radio show with human surgeon Dr. Robert White, who was transplanting the head of one monkey onto another and was on the Scientific Council of the Vatican. I was the expert witness of gross animal abuse and neglect in the notorious Silver Spring monkey trial that helped put People for the Ethical Treatment of Animals (PETA) on the map. In the UK in 1989, I witnessed, helped expose, and forced into retirement a renowned senior scientist and Fellow of the Royal Society (FRS) from the National Institute for Medical Research over his cruel and absurd experiments on cats. I confronted a university professor of psychology in the United States who was giving restrained dogs inescapable shocks in the belief that they were a model for studying learned helplessness in humans and drew attention to Psychologists for the Ethical Treatment of Animals. I visited laboratory animal research facilities, government primate research and breeding centers, and zoos to advise and lecture on environmental and social enrichment. I even had a run-in with Jane Goodall when she was visiting the HSUS, because she would not take a stand against the use of chimpanzees in research or oppose the government setting up breeding facilities, which she told me 'would help protect those in the wild.' She has subsequently changed her opinion and ethics.

It was evident to me that the denial of animal sentience, of their rights and inherent value by those involved in their exploitation, was a necessary distancing to avoid the censorship of conscience and of public censure and accountability. It paralleled those who exploit the natural environment as a mere resource with total disregard for the rights and intrinsic ecological value of indigenous plant and animal species and of sustainable, traditional native communities and cultures.

ETHICS AND CONFLICTS

The time I spent with Arctic Inuit and Native American Indian peoples and familiarity with their writings gave me insight into how our ancestors dealt with the necessity of killing animals for their own survival. I found that cross-cultural reverential respect was the hallmark of these peoples, and they had no word for pest, vermin, weed, or trophy animal.

Catholic priest Matthew Fox (no relation except in spirit) was silenced in 1989 by the Vatican police, headed by Cardinal Ratzinger (who subsequently became the next pope) for 1 year because of his appeal for a more creation- and Earth-centered ethics and spirituality, which, from a Christian perspective, called us to empathize so deeply with the life around us and within us that we receive the stigmata of an Earth crucified. The cardinal's response to the growing animal and environmental rights and justice movements at this time, which was the basis of his medieval disciplinary action against dear Matthew, was that these movements were based on 'somewhat anti-technical, somewhat anti-rational concepts of man as united to nature, and have an anti-humanist element.' Primitive paganism indeed – heretical beliefs!

It was also at this time that genetic engineering biotechnology and the patenting of life entered the global marketplace, along with genetically modified crops, farmed animals, and ever more varieties of laboratory animals to serve as models for human diseases and drug testing. I was involved at many levels in questioning these developments and extensions of our presumed dominion over other animals and the natural world, writing the appeal to the U.S. Congress to prohibit the patenting of life and addressing public forums in Canada, England, Finland, Denmark, Italy, Tanzania, and India.

I was demoted and virtually silenced by the CEO and CFO of the HSUS in 1997 because I had endorsed and written the foreword to a book critical of the pet food industry, Ann Martin's *Foods Pets Die For*. The pet food industry had complained in a formal letter to him about my endorsement. He dismissed all documented, legitimate, science-based, and clinically confirmed nutritional and diet-related concerns detailed in this book and the significant risk of many manufactured pet foods to the health of cats and dogs, many of whom belonged to supporters of the HSUS. In my defense I pointed to the HSUS's 'Eating with Conscience' campaign, which I had initiated to advise members about the suffering of factory farmed animals and that it was simply an extension of that campaign to address manufactured pet foods. I subsequently published the book *Not Fit for a Dog: The Truth about Manufactured Cat & Dog Food* (2009), co-authored with two other veterinarians, to further document these pet food health concerns.

It was a point of irony that my work and reputation had given the HSUS the scientific credibility that was needed in its advocacy for animal welfare and protection and was one of the cornerstones of its effectiveness, public recognition, and financial enrichment. My position as a vice president of the HSUS was demoted to 'Senior Scholar, Bioethics.' At that time, I was a frequent guest on Johnny Carson's the *Tonight Show* and other TV and radio programs and had many media interviews nationally and internationally, as well as issue-related speaking engagements at conferences at home and abroad. Thus, it would have been unwise for the HSUS to have fired me, since all my public engagements helped publicize the organization.

My wife Deanna Krantz and I had so rocked the boat exposing animal welfare and wildlife and environmental protection issues in India that the chairman of the Animal Welfare Board of India and the director of World Wildlife Fund, India, told my boss that I would probably be killed if I were ever to set foot back in their country. I was soon due to return and join Deanna at her animal refuge and veterinary hospital in the Nilgiris, South India. This exaggerated seed of disinformation succeeded in accomplishing its task. Then president of the HSUS announced that I could only go to India on my own time and at my own expense, implying that HSUS could not be held liable if anything were to happen to me, because I would not be on official HSUS business. Despite his betrayal and gullibility in believing this fabrication, I did return to India without incident!

SOLUTIONS AND RESOLUTIONS

Now approximately 20 years later, I see the business world and aligned industries seeking to change their public image to appear more animal and environmentally friendly. This is ultimately enlightened self-interest because when we harm the

Earth, we harm ourselves, and when we demean animals, we do no less to our own humanity, as I detailed in *Bringing Life to Ethics: Global Bioethics for a Humane Society* (2001).

After retirement from the HSUS in 2002, I was able to devote more time to assisting my wife Deanna's work in India – the self-proclaimed 'largest democracy in the world' – where she continued to run a large animal shelter and provide free veterinary services to poor communities in the South India Nilgiris Global Biosphere Reserve for nearly a decade. Her work and the animal situation in India are summarized in our book, *India's Animals: Helping the Sacred & the Suffering* (2016).

From my perspective as an advocate of the One Health concept, our own health and ultimate well-being of future generations are dependent on a healthful environment and healthy plant and animal populations and natural communities. The call for animal rights and eco-justice, too long ignored, now means that planetary CPR – conservation, protection, and restoration – needs to be immediately implemented. Our commerce with the Earth must become one of mutually enhancing relationships rather than relentless exploitation, destructive invasion, and human infestation.

Climate change, ocean acidification, loss of cultural and biological diversity, and pandemic diseases are evidence enough that democracy must become all-inclusive. 'When we take care of the earth, the earth will take care of us,' a Pennsylvania Dutch farmer once told me, adding, 'and that includes caring for the animals.'

As Dr. Schweitzer advised, the scope of ethical concern should not be limited to our own groups and species, but to embrace all living beings. As I emphasized in my book, *Inhumane Society: The American Way of Exploiting Animals*, this is ultimately enlightened self-interest, because when we demean and exploit other animals without question, we ultimately demean our own humanity: Animals can therefore help us evolve and free us from the ethical void of an increasingly depraved existence of exploitation and spiritual corruption.

A THOUGHT FOR THE DAY AND EVERY DAY

Let us embrace our gift of life and then give more than we take to live in balance and harmony in a world of takers so that our legacy honors who we were and all whom we embraced.

In our respecting and protecting other sentient beings, we elevate our own. In our loving concern, we affirm the sanctity of life, including the lives of those who are demeaned and exploited by those of our own kind, who bring ever more suffering and chaos into the world, which they justify in the name of progress and necessity.

We and all living beings are interconnected, interdependent, and interrelated in the co-evolving matrix of the Earth community, which, as my friend the late Fr. Thomas Berry asserted, is 'not a collection of objects but a communion of subjects.' Metaphysically the universal is manifest in the particular; the macrocosm in the microcosm. The One in All.

When we disrupt these connections, which some call sacred, just as when we harm other interdependent life forms in this community, we ultimately harm ourselves, setting in motion processes of devolution and dis-ease.

Beyond all that may cause despair and hopelessness and make us rage against the dying of the light, the illuming powers of humility, love, and compassion in action can help redeem humankind, recover our sanity and purpose in life, and heal the Earth and all our relations.

Index

Note: *Italic* page numbers refer to figures.

D

E

F

Q

R

S

T

U

"In this new, distinctive book, Michael Fox is covering One Health in an entirely comprehensive, easy-to-understand form. He is one of the most, if not *the* most, qualified person in the world to author a book on this collection of topics. One Health, an integrated, environmentally sound method for preserving health, is transforming the way the world is approaching healthcare on all levels and for all living species: humans, animals, plants and beyond. I am thrilled, as I know readers will be very pleased, that the author has made this major effort to write this tome, possibly his last. His writing style is outstanding and his topics riveting, making it difficult to put down! I believe this book will become a gamechanger to better understand how healthcare needs to evolve to assure a healthy planet for future generations."

–**Craig Carter**, *Professor & Director Veterinary Diagnostic Medicine Departments of Veterinary Science and College of Public Health, USA*

"Dr. Michael Fox clearly and succinctly lays out the solution to the most critical problems facing humanity, the planet, and the animals with whom we share it. At a time when reasons for optimism are rapidly disappearing, this empowering book restores the possibility of change."

–**Neal D. Barnard**, MD, FACC, *Adjunct Professor George Washington University School of Medicine President, Physicians Committee for Responsible Medicine Washington, DC*

"As the name of this book suggests, our goal must become *One* Health: human health, the health of other animals, and the health of the ecosystems that sustain life. Who better to share a vision of achieving One Health than Dr. Michael Fox whose many decades as a veterinarian, scientist, author, and newspaper columnist give him unique insights into the connections we must consider in addressing the challenges we face and the ethics we must adopt in the process. In this collection, you'll discover ways to make better individual choices while transforming systems so that they are more humane and sustainable for all life."

–**Zoe Weil**, *President of the Institute for Humane Education and Author of* The Solutionary Way *with a Foreword by Jane Goodall*

"Of all the emergent paradigms, the One Health concept is inarguably among the most important, so it is vital that one our most forward-thinking and revered scientists, Dr. Michael Fox, has produced this timely and stimulating volume. In his book, Dr. Fox sets out – as few or no other could – to integrate a vast panoply of evolutionary, medical, veterinary, anthropogenic, and many other factors. In doing so, he engagingly explains the nature, ecology, and process integral to the One Health issue. It follows, then, that this is a book for all people from any – literally any – field of interest.

Few can be left in doubt that health and disease are not simply about the battle between pathogens, immunity, and fitness, but also about the rights and wrongs of people and society, and how we address the planet on which we, and so many others, live. On many fronts, humans do not emerge well from this book, not least due to their persistent disrespect for the very home world that supports them, and our unmatched brutality to other lifeforms. Dr. Fox does not shy away from any issues, including consciousness and enlightenment, because they manifestly reside as true pillars of any progress if people are to heal their ways and the wider environment. Benevolent consciousness is a big ask of humans, but if any book can offer a rational guide towards such enlightenment, then surely it can be found within these pages."

–**Clifford Warwick**, PGDip (MedSci), PhD, CBiol, CSci, EurProBiol, FRSB
Biologist & Medical Scientist, UK

"Dr Michael Fox's new book *One Health: Veterinary, Ethical, and Environmental Perspectives* exposes and explores the loss of biodiversity on our planet and why we seriously need to take our earth and its species in a new direction. The data and facts he presents are overwhelming, but presented in a way that shows the urgency of what we need to do to be better stewards of One Health and One Earth. Climate crisis, loss of biodiversity, loss of microbial and animal species and electromagnetic field, toxins, and pesticides, as well as the overpopulation of humans taking over land, once part of natural habitat, is radically changing our earth. As Chief Seattle advised 'Whatever falls the Earth, then falls the son to the Earth. Man did not weave the web of life; he is merely a strand in it. Whatever he does to the web, he does to himself.' Zoonosis and reverse zoonosis may be the pandemics of the future. Animal agriculture has been damaging our earth in such a serious way, and Michael Fox provides the data and statistics that should make this piece of everyday life front and center in our thoughts. Bioethics should direct veterinary medicine to do the best for the animal and somehow balance the financial, corporate, and pet owners' pressures. As a predominantly plant-based, holistic human, reflected in how I feed and treat my animals, this book makes so much sense. We need to bring it to everybody!"

–**Margo Roman**, DVM, CVA, COT, CPT, MASH
Main St Animal Services of Hopkinton, USA

Made in the USA
Monee, IL
13 September 2026

62977456R00105